COMMAND SAFETY

by

Alan V. Brunacini and Nick Brunacini

Command Safety

Written by Alan V. Brunacini and Nick Brunacini

Designed and Edited by:
Uptown Graphics and Design
9143 W. Lone Cactus Drive
Peoria, AZ 85382

ISBN 0-9747534-1-6

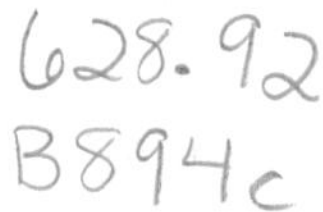

Command Safety

Acknowledgments

The material in this book is the result of the recovery project that occurred within the Phoenix Fire Department after the March 2001 death of Firefighter Bret Tarver. Bret was killed during a fire fight in an old grocery store in downtown Phoenix. After that tragic experience, the Phoenix Fire Department engaged in a recovery program to learn everything that occurred at that incident and then to develop the procedures that would improve the safety of how we conduct hazard-zone operations. As a part of that recovery process, a group of PFD officers held a series of study sessions to discuss how to more effectively connect and integrate the strategic level of incident command to the safety and survival of our firefighters. We used the eight standard functions of command as the basic framework for those discussions. Those very energetic discussions produced a set of notes that attempted to capture the content (and context) of those discussions. The material that came out of those discussions has become a part of a number of improvements to our operational and command system. That recovery process is still underway and has created an enormous amount of "developmental energy" within the PFD. I am hopeful that our dedication to Bret Tarver will produce an ongoing commitment to improving the safety of firefighters that never stops.

After our study group finished their work, I continued to refine the notes we had developed. I enlisted my son, Nick, to help with the project. He is a shift commander (deputy chief) who actually applies all the old and new procedures in the street everyday. He has added a level of realism to everything we have developed in this writing project. He along with his shift commander colleagues are also the lead instructors in our Command Training Center Program, where they actively instruct and interact with all of our department members. A major part of all that instruction includes the integration of basic safety into everything we do. Nick is a twenty-five year PFD member, an excellent writer and a very practical, no-baloney "B"-shift kind of guy. His writing is a big part of the material in this book.

As usual, I enlisted Kathi Hilmes and Harold Pickering to convert a set of expanded notes into a book. Kathi has been my PFD pardner for the past twenty-five years. She is a huge help to me and is a major player in the team who gets their crazy old fire chief through the day. She is the commander of the grammar/punctuation/editing sector and is a talented electronic graphic artist. Thank heaven, she got an "A+" in English throughout her schooling to somehow balance my "C-."

To avoid the clumsiness of frequently referring to the IC as he or she throughout the text, the author has taken literary license to replace "he or she" with "their" when referring to the IC and other non-gender specific subjects, and until a new generic pronoun without masculine or feminine connotations is devised, I will continue to use the plural pronoun with a singular subject (Kathi says I'm not in agreement with indefinite-pronoun antecedents--so I have put this in to let you know that I did it on purpose).

Phoenix Firefighter-Paramedic Harold "Pooney" Pickering has again done his usual artistic magic to cause the written stuff to come to life. His amazing artistic ability and street experience create a graphic gift that makes words more understandable to read, and a lot more fun to look at.

Our old friend, Doug Forsman, has also done his usual excellent job of coordinating all the different pieces and parts that must be somehow shoved in one end so they can come out the other as a book. He combines being a fire chief (Greely, Colorado) with an excellent knowledge of publishing and production, and is always able to make the process fun. He has effectively connected with everybody and everything at just the right time. The Oklahoma State University-IFSTA team are always a huge help to us. It is truly a joy for us to deal with and hang out with Chris Neal, Janet Maker, and Mike Weider. Their support and assistance simply make the entire Fire Command adventure possible. It has been an honor for me to have been connected to Oklahoma State University, starting as a young Fire Protection student (class of 1960) to now lapsing into old age with them. OSU has been an important part of everything I have been able to do in my fire service career and I am eternally grateful for their kindness.

The very basic objective of this book is to improve the safety of the firefighters who must go into the hazard zone and do the most important part of our business. They provide the brains and the guts that create the ability to go into those places. The incident commanders who are charged with the responsibility to command and manage those operations must also have the brains and guts to create the strategic support that causes those firefighters to always get out of those hazardous places. We hope this material is a help to both the firefighters and the IC.

AVB--2004

Command Safety

Table of Contents

Th... ety and survival

o... who fight on

...rver.

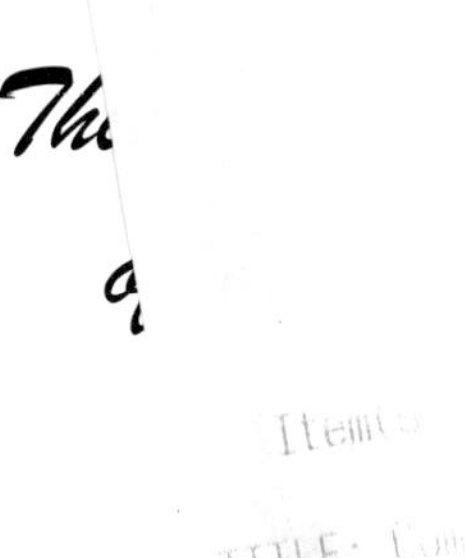

Printed with permission from Steve Benson, Arizona Republic.

Command Safety

Up-Front Stuff

Command Safety

Up-Front Stuff

The death and injury of firefighters continues to be the most serious challenge facing the American fire service. Many of these occupational (i.e., tactical) fatalities and severe injuries occur during structural fire fighting operations. Putting fires out in buildings is the original mission of our service and the most enduring traditional activity we engage in. Structural fire fighting continues to also be the fastest and most violent activity we engage in, and it is clearly the most consistently dangerous thing we do. A major contributing factor in almost every firefighter death that occurs on the fireground is some difficulty with incident command. These problems are the result of too little, too late, not correct, or somehow out-of-balance application of the command system. The person we call upon to initiate, manage, and expand (as necessary) the incident management system (IMS) is the incident commander (IC). That person must play the command and control role for the overall incident. The IC completes their operational and safety responsibility to both the customer and the firefighters, by performing the eight standard command functions. These functions serve as the very practical (simple, understandable, doable, reviewable, enforceable) performance foundation for how the IC completes their responsibility, as the strategic-level incident manager, and the overall incident safety manager.

How the IC manages the safety of the troops, who conduct hazard-zone operations, is a major challenge that occurs at every incident. The command system calls upon the IC to manage and control the strategic level of both operations and safety. A major objective of the IMS is to create, support, and integrate operational and command bosses, who "cover" the geographic and functional needs of the entire incident on the task, tactical, and strategic levels. Huge safety problems occur when these three (standard) levels are not in place, operating, and effectively connected. It is impossible to achieve an overall ("regular") level of safety management, until and unless an IC quickly gets set up on the strategic level at the very beginning of the event. The IC must then build an effective incident organization to create the capability to stay on that strategic level, as long as firefighters are operating in the hazard zone. The faster the system can get a strategic-level IC, the safer hazard-zone operations become.

Command-level safety is always the IC's most important responsibility, and must be automatically integrated into the regular command functions. Simply, we can't take care of Mrs. Smith, if we are trying to save ourselves. Our customers call us because they and their possessions are threatened by physical, social, and medical conditions/hazards that they cannot control themselves. We are the local, home delivery, up close and personal, hazard-control agency that has a three-digit phone number (9-1-1), so we are easy to call. What local means is that we are the agency that actually answers Mrs. Smith's 9-1-1 call, and then translates that call into an immediate response, that directly and physically delivers service to the Smith family. There is virtually no other agency with that consistently short-response time local service delivery system capability in place... if you want the army, you've got to have a war; if you want a "red card" Federal response, you've got to have a huge disaster; if you want a plumber, you've got to have a week--we are "blue card" hometown

responders who show up in four minutes... that's local. To stay in the local emergency service business, we must create and maintain the ongoing, day-in, day-out (24/7) capability to quickly enter the hazard zone, do the work that is required to protect the customer and their stuff, and to then consistently be able to exit the hazard zone (i.e., make a "round trip"). Being able to conduct effective, safe, hazard-zone operations is what defines us as emergency response professionals more than any other single thing.

We make a promise when we become firefighters that we will quickly (and safely) respond when the customer calls us for help. That promise also includes that we will put our bodies directly in between the incident problem and the customer. This is how we actually (physically) protect the customer. Our job is to always be effectively prepared and highly persuaded to aggressively conduct the local (blue card) "street fight" that makes Mrs. Smith's problem go away. The IC must create and manage a system (evaluate/decide/order/act/revise/survive) that closely and continuously connects command and safety to protect us, so that we can both do our job (i.e., protect the customer), and live another day (hopefully, much longer). The safety system and the command system must be automatically integrated and done together. The IC typically does not have the time nor the attention span to do these two separate major routines (command/safety) under fast-and-dirty, local incident conditions.

Command Safety

Up-Front Stuff

The foundation of our safety/survival mentality is that we think (and feel) that our members have a permanent connection to our organization, to what we do, and to what we stand for. Simply, we are all "lifers"... the only thing we keep for thirty to forty years is our humans; therefore, it makes intellectual, emotional, and "business" sense to always package them up, and protect them in a system that operates and manages in a way that causes those humans to be both effective and safe, while they do their jobs. Our safety approach is just as simple as the difference between how I treat a rental car, and how I treat my over-restored '72 Chevy pick up--I have never changed the oil in an Avis or Hertz rental (I wipe off the engine in my '72 every three days.

In addition to the promise we make to the customer, we also make a promise to each other that we will always take care of each other. This promise applies particularly to how we protect each other during hazard-zone operations. These promises are absolutely necessary for both Mrs. Smith and for us. In order for Mrs. Smith to get through the day-to-day wear and tear of living her life, she must have the "quiet confidence" that we will always be there, when she need us to protect her. Sometimes things go wrong when she is frying French fries and the phone rings in the living room (with blabber-mouth Mabel on the other end), driving her car through modern, maniac-infested traffic(!), or when she tucks little Suzie in at night, with what she thought was a case of sniffles, and forty-five minutes later ends up with a "I can't breath, Mommy" situation. Then, she must depend on E-1 quickly showing up and making the problem go away. The guys and gals on E-1 must also have the same deal. They must operate with the confidence that when they go into the hot zone to get the Smith family out (for whatever reason), that everyone on their team is on the same page of the standard operating procedures (SOPs), that everyone on the team agrees to play their assigned part in the standard incident operation, and that everyone will stay connected to all the other levels. They also promise (just like to the customer) that they will put their body in between the incident hazard and their fellow firefighter to protect that person. This internal agreement creates a strong set of feelings about each other. This internal agreement also creates powerful instincts to protect each other when we are threatened ("you go--we go"). Based on these strong cultural (protective) emotions, the most disruptive and tragic thing that can occur to us is for one of us to be trapped, injured, missing, or killed. When this occurs, we instantly and completely redirect and concentrate our attention, energy, and resources to rescue the members of our team. These desperate recovery events absolutely destroy our regular service delivery routine (along with the systems we use to manage those events), and quickly become the most emotional, confusing, and horrible experiences that can occur to us.

This built-in culture of self protection creates a strong reason for us to always operate in a standard and safe manner. We are in business to hop on big red trucks (BRTs), and charge off to Mrs. Smith's when she calls us, because her kitchen is on fire--not to run a complicated, internal safety program. In fact, she (customer) expects us to know what we are doing when she needs us, and for us to effectively protect ourselves while we do our jobs. A basic service-delivery reality (recurring in this essay) is that we can't protect Mrs. Smith if we can't protect ourselves, or if we are trying to save one of our responders, because that person is in trouble and now must be rescued. Our safety program creates the basic foundation for effective service delivery. If we don't have and practice such a standard safety program, we are (in actual effect) just transporting more potential fire victims to the incident on fire apparatus. These operational realities create a tough safety job for the IC. Mrs. Smith depends on "her firefighters" to be there when she needs help and to solve her problem. That problem can produce a high-hazard situation that requires strong, fast, hot-zone action. "Her firefighters" are (thankfully) highly motivated to directly, personally, and physically do whatever is required to help her. Fast, effective performance on kitchen-fire day is what they feel is the most important part of their job. Those feelings are a good thing. As long as acting them out occurs in a standard manner, and safe is always a regular and important part of standard.

Based on both the inside promise (to each other) and the outside promise (to Mrs. Smith), a major part of our safety program must first focus on and provide active support to do whatever is required to prevent us from getting in situations that require high-risk rescues of each other. The system must also describe in detail how such rescues (i.e., responders saving responders) will be conducted, when they are necessary. The entire safety system must focus on what it takes to prevent such rescues, because of how difficult, desperate, and dangerous they are. The very best way to protect our members is to predict, manage, eliminate, or avoid the situations/conditions that require us to have to save each other. Serious, sober, safety students should reread the last sentence. No matter how much we love fighting fire, and all the exciting things that go with it, the absolute worst thing any firefighter will ever endure is a situation where one of us is in trouble inside the hazard zone, and now we must expose more of our humans (generally a lot more) to those same many-times deteriorating fatal conditions, to save our firefighters. This is where/when the romance of fire fighting instantly turns into the terror

of fire fighting. Everyone operating on the fireground must understand that the success of these high-risk rescues is very low. We must never let routinely operating in and around hazard zones, where those dangerous conditions become so familiar to us, that we become anesthetized (unconsciously detached) to what the word (and the reality) HAZARD really means. When hazard-zone conditions have "grabbed" one of us, it is many times virtually impossible to locate and remove that member in time to save them. The sad conclusion of these awful responder-rescue events, many times, involves the horrible sound of a body bag getting zipped up.

The IC must effectively combine our obligation to act with the standard safety activities that are required to keep the troops alive and okay. The basis of this approach involves the consistent application of a standard safety plan. The foundation of how we must operate is the application of a short, simple, easy to understand (not so easy to actually apply) risk management plan. This plan is based on the IC doing a continuous assessment of the capability of our standard safety system to protect the firefighters from the hazards that are present. The following three-level plan should serve as the basis for us conducting hazard-zone operations:

1. We will risk our lives a lot (in a highly calculated manner) to protect a savable life.

2. We will risk our lives a little (in a highly calculated manner) to protect savable property.

3. We will not risk our lives at all in an attempt to protect lives and property that are already lost.

The IC must approach incident operations with the attitude that the fire and the building are always trying to first assault, and then assassinate our firefighters. In order to protect our troops, the IC must always connect standard action, with standard conditions, to produce a standard outcome. A realistic evaluation of the current incident situation, combined with a thoughtful forecast (based on past experiences against a standard scale of incident stages) will serve as the most accurate prediction of where (on the scale) the incident is now, and where it will ultimately end up. Simply, the middle and end of the incident must emerge out of the beginning (where else would it come from?). The IC can't unburn what has already burned, resurrect those who are already dead, or somehow shove the explosion back into the tank. The IC must use the regular pieces of the regular safety system to create a standard operational response that protects the troops from the overall incident hazards. In offensive situations, the safety and tactical response must be big enough to adequately protect inside operations. The IC must always consider the profile of the "round trip" (in and then out) in offensive situations, where conditions will allow firefighters to make an inside attack. If there is any question about a safe exit, the IC must either increase the support to the inside attack ("harden the exit"), or redefine the strategy as defensive.

This "round-trip" approach creates a more modern (and a lot smarter) definition of what an offensive situation actually looks like. We used to say (in the "old days") that if attackers could get in, they should go in (period). That was what "offensive" meant then. Now we are closely and carefully attaching an exit to every entry. The IC and the whole team must evaluate the entry/exit ("round trip") profile of a hazard zone, before they order and extend an offensive attack. The most dangerous situations are those that produce conditions that invite you in, and then quickly get you in their clutches, first assaulting you, and then assassinating you. The standard safety response also must always separate the firefighters from defensive conditions (collapse/heat/toxins), so they don't get those conditions "on them," because once you get defensive conditions on you, it's just about impossible to get them off. What defensive means (simply) is that our regular safety system is not capable of protecting our firefighters, so those troops must be moved away from (and then kept away from) those excessively dangerous (in comparison to our safety system) conditions, until those conditions are either put out or burned out.

Throughout this essay, we will refer to the balance between the level of hazards that are present in relation to the size of our standard safety system. We call this relationship the "safety math" and this equation becomes a simple, fast-and-dirty way to make operational position/action decisions, about the survivability of where we can go, and what we can do. This index of survivability becomes the very practical foundation for the IC making the offensive/defensive strategy decision for the incident. The IC cannot assign firefighters to positions where the safety system will not effectively protect those firefighters. The process is not complex: if the safety system is bigger than the hazard, we can go there/if the hazard is bigger than the safety system, we can't go there.

The basic, structural fire-fighting incident hazards (that consistently injure/kill us) involve:

- ☐ structural collapse
- ☐ toxic insult
- ☐ thermal insult
- ☐ trapped/lost.

The basic system components that we use to protect firefighters include:

- ☐ adequate number of trained, fit firefighters
- ☐ personal protective equipment (PPE)
- ☐ operational hardware and water
- ☐ safety SOPs
- ☐ IMS.

The beginning point of our safety plan must be that every team member takes responsibility for their own compliance with the department's safety plan. This personal responsibility requires every individual to always behave in the following manner:

- ☐ understands the details, dynamics, and effects of the basic fire-fighting hazards

- ☐ understands the details of the department safety plan and the limitations and capabilities of the safety system components that go with every task, and automatically follow the safety plan

- ☐ is physically, mentally, emotionally, and organizationally capable of executing the details of the safety procedures for their position (fit for duty)

- ☐ monitors their own safety and welfare

- ☐ continually evaluates and self adjusts their own safety procedure compliance in relation to the incident hazards
- ☐ directly stops any unsafe acts they can impact
- ☐ always assists those close to them
- ☐ actively reports safety conditions, status, and changes up/down/across the incident organization
- ☐ quickly cooperates and complies with safety orders and instructions.

The IC must operate with the expectation that everyone is voluntarily and automatically following the regular safety plan. If this does not occur, we lose the capability to conduct standard incident operations, and the incident becomes a lot more like a free enterprise rodeo, than an organized incident operation. Effective incident operations require officers on every level to act like real, live leaders who play their standard operational and safety position. Everyday safety compliance occurs because of good SOPs, training, understanding, skill, system agreement, personal cooperation, direction, and practice. This safety survival process must be effectively in place, ahead of the incident. Show time (fire fighting) is a lousy time to introduce the basic safety procedures to the players, who must now use that system to protect both the customers and themselves. Bosses, on every level, must continually commend those who follow safety SOPs. This positive approach becomes the most powerful way we reinforce

The safe way must become the only way.

those procedures, and make that consistent and automatic safety mentality and response, individually and organizationally (i.e., collectively), a habit. Responders (on every level) typically want to please their boss and to do a good job. Having your boss tell you that you did a good job, and that they appreciate you following the safety procedures is a big deal. If there are responders who don't understand the basic safety routine, they should be sent back to safety school. If they forgot the details of the safety story, their boss should review the basic safety routine with them. If their noncompliance is a function of a misbehavior problem (as opposed to a performance problem), or if they are suffering from some strange noncompliance personality disorder, they must be put in "time out," and have some organizational intervention extended to them. If their misbehavior consistently outperforms the (behavior) recovery capability of the organization, they must not be allowed to operate in the hazard zone. Bosses simply cannot tolerate safety whackos, because of the potentially fatal consequences of their non-compliant behaviors to themselves and others. Safe and effective operations are just that--safe and effective. In the end, the safest way to do something is also the quickest and most effective, and must become the only way we perform.

A lot of our non-safety SOPs are written as guidelines, where the procedure is used to basically establish effective boundaries that help us hang around the center line... this (center line) is what "standard" really means. Everyone should be empowered to use the regular, non-safety SOPs as a starting point to deliver basic, effective core service, and then to customize an added-value response to meet the customer's special needs. As a contrast to "regular" SOPs, safety directives must be considered and applied as rules (i.e., commandments) that are strictly and consistently enforced. Simply, no one is empowered to break a safety rule. In fact, being safe becomes the (modern) beginning for authentic, effective, organizational empowerment. The basis of empowerment is the creativity of our members. This is a good thing, and generally produces a positive service-delivery response, that leaves the customer with a very positive memory of their fire department. Creativity does not apply to safety rules. Whatever the rule says, we always do, and we always do it, every time. We automatically follow the safety rule if we are alone, or if we are part of a response mob. We all understand the safety rules, and do not need anyone to teach them to us again, to remind us of them, or to force us to do them. This voluntary compliance process requires smart, committed adults who seriously understand, follow, and support the safety SOPs--no matter what position they play. The operational safety challenge should involve protecting ourselves from the conditions created by the incident problems (which is why Mrs. Smith called us in the first place), not that we have responders who are blasting right past the regular safety routine, and are doing Evel Knievel daredevil imitations. Every responder must operate with the confidence that every other responder is performing within their assigned role, and is following the basic personal

Command Safety

Up-Front Stuff

and positional safety SOPs. Inside the organization, everyone is assigned a standard operational and safety job that automatically goes with that position/job/level. Every officer is responsible for safety direction and supervision on their level, and for the safety, survival, and welfare of everyone under their command. Safety must be automatically built into performing the regular functions of the strategic, tactical, and task levels. Every level must effectively connect and integrate with the other two levels. The responsibility for safety increases up the chain of command (given the organizational "pyramid," the IC is sitting on top). Complete safety coverage occurs only when the three levels are in place, operating, and connected to each other. The basic three-level safety approach is arranged in the following manner:

LEVEL	OFFICER	SAFETY RESPONSIBILITY	VIEW ORIENTATION
strategic	IC (sitting boss)	all incident areas/ functions	overall (entire incident)
tactical	sector (walking boss)	geographic area/ special function	medium (their geographic/ functional assignment)
task	company (working boss)	their direct work supervision	short/local (their assigned work location)

While the three levels must be closely connected by the entire command system, they are physically separated when the incident operations and command "game" starts, and each level must be effectively in place, so they can perform their assigned part in the safety plan. To be consistently effective, all three levels must be "positioned" to do their job on their own assigned position, function, and level. While the three levels must assist and support each other, one level cannot consistently make up for the nonperformance of one (or both) of the other levels. The basic safety system must always be implemented and prepared to quickly respond to help (i.e., rescue) anyone (on any level) who is in trouble, and must be automatically escalated to a point, where the size of the safety system matches (or exceeds) the incident level, and the relative size, complexity, and duration of the hazard that goes with that level. Such rescue responses of our own troops should be caused by the effect of unseen or rapidly changing conditions, not that standard safety procedures are not being followed. What we do (because of where we must go) is highly hazardous, even if we follow all the safety SOPs... it's eventually suicidal, if we don't. We say "eventually" because we can get away with breaking basic safety rules for a long time. We have smart, aggressive workers (i.e., firefighters) who are highly system savvy, and can be the best rule breakers on the planet. Sometimes, they practice and refine playing fast and loose around hazards because it's exciting and fun. This ongoing rule breaking causes us to lose track of how dangerous incident operations can be, and pretty soon we begin to lose respect for the hazard zone (big mistake!). The problem is that such habitual behavior makes us dumb, and on a dark and windy night, the "note comes due," and our well-practiced unsafe habits get us into a spot we can't get out of, and now incident conditions assault/assassinate us. Bosses on every level must continually and critically review ongoing safety practices/habits for complete SOP compliance, and must not let sloppy performance or close calls go unattended and uncorrected. Based on how important and connected safety performance is to overall operational

effectiveness, we must front-end load every level with the ability (SOPs/training/application/review/revision) for that level (person/place/role) to do their assigned operational and safety job. Performance for each level must be managed, and actually occur, specifically on (i.e., within) that level, and this is how we "cover" thc entire incident with effective safety management.

Virtually every fire service boss started as an entry-level firefighter "grunt," and based on the powerful socialization of all of our initial-task orientation, when there is a firefighter accident/injury/death, we naturally resort to developing a solution that connects us to that strong, initial, task-level socialization. Many times, that task-level response does not match where the problem is actually occurring. Our habitual response causes us to focus on the task level where the problem (injury/death) generally occurrs. We then almost automatically resort (default) to creating a response that instinctively goes "back to basics" on the task level. The problem that actually caused the injury/death many times actually started on the strategic/tactical level, so this almost automatic task-level recovery response causes us to misplace our recovery efforts, and we end up doing the wrong thing harder. Many times this involves us sending already well-trained firefighters back to hose and ladder school... while basic training (on any level) is a good thing, in this case, this task-level training will not prevent safety problems that exist on another level... simply, sending workers back to worker school won't solve boss-level (IC) problems.

Note:

This modest essay is an attempt to describe a standard, hazard-zone, safety management plan for the strategic-level IC. Hopefully, this material could be used as the basis for a training/study program, a very practical operational manual, and then as an outline to critique the performance of an IC after an actual incident. It's pretty easy to criticize how an IC managed the welfare and survival of hazard-zone workers. It's not so easy to describe (in detail) what the IC should do to complete their part in that process. Our ICs deserve to be well trained and to have the opportunity to understand and practice how they should properly perform their strategic-level job. The survival of our troops depends on us doing this.

By not addressing the problem right where it is occurring, we continue to make the same mistakes that injure/kill firefighters at incidents where the problem really was that the IC was not effectively performing the standard command functions, not that the firefighter couldn't do the task-level "basics." A guy with a wild hairdo (Einstein) once said: "Insanity is doing the same thing over and over, and expecting a different result"--this about sums up what is going to occur, unless we can add a standard, strategic and tactical level of safety support to our (instinctive) traditional task level. Anyone who doesn't believe it, should look at the firefighter fatality statistics... the problem marches on, and will continue to march on until we get serious about a three-level command and safety system, where every level must first do the command, operational, and safety routine for their own level, and then effectively stay connected to support the other two levels.

Early in our careers, we quickly develop a strong connection to playing our part in task-level fire fighting operations. Being able to effectively perform "when the chips are down" is how we become a legitimate member of our fire company. This initial "everyone starts on the bottom" entry and assimilation process predictably causes us to

develop a strong emotional attraction to the physical work that is involved in fire fighting. The work is typically very direct, dirty, and dangerous (i.e., up close and personal). Being able to effectively (and aggressively) do our part, in physically doing this work, is how we are accepted as a firefighter, and this becomes the basis for us moving up the promotional ladder, and this capability is the very powerful foundation for the peer process.

The early attraction (and attachment) we develop to the action-oriented, physical labor part of our business serves us very well in the beginning of our careers. This initial socialization and practical orientation should serve us very well throughout our careers (if we are well adjusted). Actually performing that work on the task level, as a fire company team member, is really the only effective way to learn what it takes to do that work. Understanding the performance dynamics of coordinated fire attack from the "business end" (i.e., fire company) becomes the very practical (promotional) foundation for both the tactical (sector) and strategic (IC) level of command operation. We are basically an organization that can quickly deliver and mobilize well-trained teams of workers, who are able and willing to do highly-skilled physical labor, that is integrated, coordinated, and managed to be effective and survivable under dangerous conditions... simply, structural fire fighting is the smartest form of manual labor. The command system is basically in place to assign and support fire companies, in hazard-zone operating positions that physically solve the incident problem. Those who are assigned as command bosses must understand what works and what doesn't work, right where the task level of effort must occur, if we are going to consistently put the fire out, and go home okay. Simply, most firefighters (including the authors) would not be comfortable (or willing) to go inside the hazard zone of an active incident (or really any incident), being managed on the outside by a bright, twenty-five year old, recent MBA graduate, who was hired to manage incident operations because of an exceptionally grade-point average. In fact, if the whiz kid is really bright, they would also not be comfortable (or willing) to be put in that position--if they were (both comfortable and willing) to take on managing an underway incident, they would be truly dangerous.

While our front-end lessons provide a robust (functional/exciting) career beginning, the enormously strong attraction we develop for the direct-action (rough and tumble) part of our job can set us up for a difficult transition later on. The problem occurs when we must recruit our members (generally officers) to serve as the IC, and as geographic and functional sector officers. The system calls upon these responders, based on assignment, arrival sequence, or standard procedure to "leave" the nozzle/hook and become part of the incident command team. They (and the system) will be effective to the extent they (and the peer process) can effectively make that (working ➡ supervising ➡ commanding) transition. The power of ongoing organizational socialization and acceptance occurs on the peer level. We all have a strong need to be part of our group. That group (peers) has the power to accept or reject us. When we get promoted, the same thing happens. The group has the power to say that we remember where we came from, and are "okay," or that we have somehow changed, and are now "not okay" (understated). When we go from the action oriented, task-worker level (doing) to the more tactical- and strategic-management level (managing), we will only be accepted and supported if the peer group approves of our past/current behavior.

What this means (in IMS terms) is really pretty simple. The peer group gets to define what (and who) is acceptable and what (and who) is not acceptable. If that peer process represents (feels, says, and acts out) that the only acceptable place for any responder to be (regardless of rank, role, or function) is directly inside the hazard zone, then any fire department member (on any level) who is outside that hot zone is unacceptable... the "real men" are inside gettin' their asses kicked, and the cold-zone IC wimps are safely ensconced in a comfortable Suburban wearing their designer command vests with the heater/air conditioner blowing across their tactical work sheet (TWS). Until and unless the authentic (not advertised) cultural dynamics of the organization will mature, develop, and allow it, it will be unacceptable in the real world for an IC to assume (and stay in) a strategic command position, so we will continue to have every responder on every level, within twelve feet of the nozzle. When this occurs, it is absolutely impossible to operate a three-level incident management program. If we do not have the personal capability and system support required to make that shift from action to command, we simply cannot (actually) create a strategic and tactical level of command. What we end up with is (in effect) a disoriented group of officers who are stuck somewhere in between a worker and a boss, and are trying to somehow create a command post (CP) on the strategic level in the hazard zone (physically located in the tactical/task area). All the command-level job descriptions, elaborate incident organizations, and organizational blab about a tactical and strategic level of command is pretty much useless, unless we can get real-live firefighters, who have been validated on the task level and then are trained, prepared, and "given permission" (by the task level) to move up to actually fill the standard, tactical- and strategic-level roles, that are required to create and operate a complete command and safety system. If all of the people whom we expect to manage the incident are emotionally, mentally, socially, physically, and directly involved in the fire fight, what we are doing (in actual effect) is sending nozzle men/women to the incident in chief's cars. Helping those officers make this national transition requires a level of organizational support that can effectively compete in a smart and complimenting not contradicting way, with their initial socialization. That process must involve the development of effective command procedures, lots of practical, simulation-based training, the opportunity to use new command skills in a supportive environment, and positive reinforcement and coaching by bosses who are creditable with the troops. That progressive command transition process must use and take advantage of the IC's basic task-level knowledge and their tactical capability, as the foundation for making this change. The IMS calls upon us to physically leave the nozzle when we are assigned as the IC or a sector boss, but not to forget everything we learned while we were there (be careful of those who have such amnesia). The command team must always understand the actual, on-line (possibly awful) dynamics of the conditions that exist where the manual labor of the incident is occurring--they must understand how the work actually gets done, and they must "respect the task."

During active and expanding incidents, it is impossible to create any sort of effective management response (both for operations

THE STRATEGIC LEVEL SAFETY ROUTINE

and for safety) when we cannot quickly produce a local "blue card" command team, operating in standard command positions. We must create and maintain an operational and command safety system that gets set up routinely, on the three standard levels. The following outline describes what happens when the three levels are not effectively in place:

- ☐ If the IC does not do their part in the safety plan:

  - There is no overall strategic-level command direction and protection for the entire incident... simply the IC is not evaluating, deciding, creating, controlling, maintaining, and revising (as required) the overall offensive/defensive strategic operational mode. Having the IC manage the incident strategy becomes the very practical foundation for a safe and standard incident action plan (IAP), and then controlling the operating positions that go with the strategic and operational approach.

  - The IC does not continually do the ongoing "safety math" to evaluate the relationship, between the size and dynamics of the incident hazard, and the size and effectiveness of our safety system, to determine the overall operational strategy.

  - There is no overall real-time, on-line (dynamic) assignment awareness, inventory, tracking or control of where fire fighting crews are actually assigned, what they are doing in relation to dangerous conditions, and if they are okay.

Command Safety

Up-Front Stuff

- There is no effective ongoing info exchange, recording, processing, and strategic management being done for the overall incident.
- The IC is disconnected from the task/tactical operational level so it quickly becomes impossible to react to changing conditions--now the command system does not have the capability to quickly move firefighters when they are threatened.
- Sometimes firefighters end up in offensive positions under defensive conditions--this can be fatal.

What basically kills firefighters at structural fires is that:

BASED ON INCIDENT CONDITIONS, THEY ARE INSIDE WHEN THEY SHOULD BE OUTSIDE!

☐ If sectors do not do their part in the safety plan:

- There is no effective on-site coordination/direction of hazard-zone fire companies = unsafe, disconnected tactical activities occur (fire companies don't have a good, up-close and personal boss or, sometimes any boss assigned to look out for them).
- There is no effective connection or integration with other sectors (no commo or info exchange)--sometimes sectors beat each other up.

Command Safety

Up-Front Stuff

- There is no effective evaluation, sending, receiving, or reacting to critical on-line/real time hazard-zone information that is exchanged among sectors and the IC--lots of dangerous surprises (IC/sectors miss identifying and reacting to the safety "red flags" in time to protect and save firefighters).

- There is no effective reporting of sector conditions, up to the stationary "sitting boss" IC from mobile "walking boss" sector officers, who are assigned and in place to be the stationary/remote IC's eyes and ears--without sector boss support, the IC becomes an isolated "mushroom" who becomes disconnected from the event, and many times does one of the following:

    1. IC stays in the CP and becomes a disconnected, uninformed, disoriented spectator instead of a critical player, or

    2. IC gets out of the CP to go see (directly) for themselves, and now becomes a distracted, hyperventilating, orbiting track star, rather than a strategic-level IC... both suck.

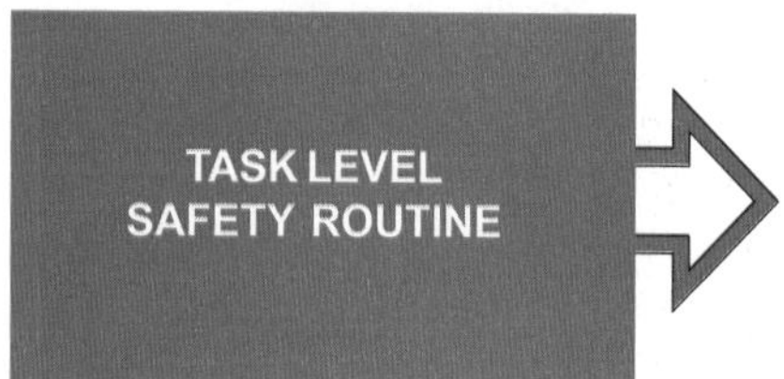

☐ If companies do not do their part in the safety plan:

- There is no effective, initial, and ongoing evaluation of immediate, up-close and personal operating position safety by the company officer... no in place, local prediction of where current, hazard-zone conditions are going.

Command Safety

Up-Front Stuff

- Companies do not follow initial accountability procedures and auto assign themselves (free-lance) based on "emotional motion" instead of rational evaluation and regular systematic (i.e., SOP) assignment.
- Based on free-lanced self assignment, no one knows where the company is, what they are doing, and if they are okay--now it is virtually impossible for the IC to effectively manage and move that company.
- There is no fire company reporting (up/across) on operating conditions and changes in those conditions (more IC "mushroom" stuff occurs).
- Companies engage in improper and unsafe operations in dangerous, unsupported positions.
- Companies free-lance themselves into positions that range from early to midpoint offensive; these conditions can (and sometimes do) quickly change, suddenly become defensive, and overwhelm safety system capabilities... this is mostly why firefighters get beaten up, and sadly sometimes die on the fireground.

The IC must monitor and react to offensive situations that quickly change and go defensive. This is when having an effective local command system, routinely practiced, automatically implemented, and in place from the very beginning of operations quickly saves us and where the exact opposite occurs when it's not in place... we use the word quickly simply because the system must be in place before there is a need to rescue one of our own--it's way too late to set the system up after we get into trouble.

Being out of balance in some way on any of the three organization levels produces the consistent, fairly timeless top five causes of firefighter deaths at structural fires:

1. lack of incident command
2. inadequate risk assessment
3. lack of firefighter accountability
4. inadequate commo
5. lack of safety SOPs.

Command Safety

Up-Front Stuff

- ☐ Huge safety/survival problems occur when there is not an effective IC in place, performing the standard functions of command:

  - Effective overall incident safety must start at the top of the incident organization with the IC.

  - The overall incident strategy creates the very practical foundation for the basic system we use, to manage the position and function of the task-level troops:

    - ➡ offensive--firefighters go inside, put the fire out on the inside, and are able to safely exit the hazard zone.

    - ➡ defensive--firefighters proactively identify defensive conditions, stay outside, operate firestreams (big ones) into and onto the involved area to limit fire spread and protect exposures; building generally falls down and goes boom, when it hits the ground; everyone who is outside the collapse zone lives to fight another day (yeah!)--two days later, DeWorth Brown (old time Phoenix demolition contractor) hauls rubble to 27th Avenue landfill; six months later, Mr. Smith completes new building--all firefighters are now six months older (alive)... life happily goes on.

  - While the tactical and task levels must actively do their part in the safety plan, the IC is the only person/place/process who can establish, maintain, and revise the overall offensive/defensive strategy for the whole incident... simply, no one else is in a position to be responsible for, or consistently able to "add up," process, and then react to what is going on all over the incident. The IC does this to continually maintain the "safety score" (safety system vs. incident hazard). When the hazard numbers exceed the safety system numbers, the IC moves the troops to a defensive position, and manages the incident as a defensive event.

- The management of the overall offensive/defensive strategy for the entire incident is the responsibility of the IC--task and tactical levels must each manage their area and the safety of their troops (and must quickly push critical safety information in their area/function into the CP), but they cannot make up for an ineffective, out-of-control, or nonexistent IC.

- The IC must automatically perform the basic strategic-level safety routine by doing the standard, eight command functions:

  1. Assumption/Confirmation/Positioning
  2. Situation Evaluation
  3. Communications
  4. Deployment
  5. Strategy/Incident Action Planning
  6. Organization
  7. Review/Revision
  8. Transfer/Continuation/Termination

☐ Integrating Safety and Command

It's easy (as hell) to blab about a lack of IMS as a big safety deal, but what do we actually do about it?... connecting safety and command becomes the very practical answer.

Command Safety

Up-Front Stuff

Integrating safety and command is the only consistently effective way to cause a functional level of strategic safety to always occur. We must make "safety" and "command" interchangeable, absolutely connected, and automatically done together, all the time/every time. Simply, the IC doesn't have the time to do a separate safety "add on" under typical incident conditions (confused, compressed, and dangerous).

The marriage of command and safety causes worker welfare to become the basis of effective operations, and creates an understandable, non-mysterious, teachable, doable IC safety routine (that no one else is in a position to do).

This manual (*Command Safety*) is a part of the *Fire Command* series. It is written around the content and organization of the *Fire Command* textbook and is designed to be a companion document to that book. *Fire Command* is written around and presents the basic details of the eight standard functions of command. The eight functions of command check lists that appear in the back of each chapter in the *Fire Command* book serve as the outline for the material presented in *Command Safety.* That material is designed and directed to describe how the IC must use the regular, everyday command functions to complete the strategic-level safety responsibilities during incident operations. Using the regular command functions creates an effective way, and a close connection between incident safety and incident command, and attempts to show how critical it becomes to worker safety, that these two activities are quickly, automatically, and effectively performed and integrated. *Command Safety* presents a detailed description of how the IC must use the local command system, as the basic foundation for doing the strategic-level safety job. Every IC must always consider the safety effect that every decision, order, and action they create will have on the welfare of the hazard-zone workers.

Incident command problems are a major contributing cause in virtually every firefighter death that occurs at a structural fire. It's pretty easy to point out problems after they occur (woulda', shoulda', coulda'). It's pretty tough for us to consistently do something proactively to prevent those problems. The material in this manual attempts to present a detailed description of how ICs can actually perform the basic command safety function to prevent the occupational deaths of firefighters.

A major focus of all the words in this manual is that our service has suffered through the ongoing deaths of our firefighters for the past 250 years. We basically have been insulted (injured/killed) by the same factors for that entire period. Simply, there are no new fatality categories. Each death comes with it's own pain, anger, and guilt. When this occurs, some fireground boss must now live the rest of their days wishing "they had only ___________." This modest essay is directed toward the authors listening to (and having) those same painful regrets throughout their fire service careers.

RNB and AVB--2004

Command Safety

Assumption, Confirmation, & Positioning of Command

Chapter 1

Command Safety

Assumption, Confirmation, & Positioning of Command

Command Safety

Assumption, Confirmation, & Positioning of Command

Function 1

Assumption, Confirmation, & Positioning of Command

Major Goal

To quickly establish and confirm a single IC and to place that IC in the most effective initial command position.

IC Checklist:

- ☐ First arriver must quickly assume initial command per SOPs.
- ☐ Use strong, quick, automatic command assumption to eliminate any command zero impact period (ZIP).
- ☐ Confirm command assumption with a standard initial radio report (IRR).
- ☐ Use occupancy/location to name command.
- ☐ Select the proper command mode (investigative/fast-action/command).
- ☐ Correctly position command to match and support the current command mode.
- ☐ Set up a standard command post as quickly as possible.
- ☐ Begin to “package” command for ongoing operation and escalation.
- ☐ Correctly accept/continue/transfer command.

Command Safety

Assumption, Confirmation, & Positioning of Command

IC Checklist:

☐ First arriver must quickly assume initial command per SOPs.

Safety Effect:

Fixing responsibility on the initial arriver to automatically assume command (i.e., become IC #1) from the very beginning of our arrival creates the best chance we have to start the operation out under control, and then to provide the ongoing command and operational capability to stay in control, and never lose control. Establishing command at the very beginning of our arrival provides the basis for establishing standard operations where firefighters can work in a safe, survivable manner that is being managed by an in-place IC.

Early incident command system (ICS) development changed the way that initial command was established. Before ICS, it was common practice that only an officer could (or was allowed to) assume command, and in many cases and places, that person had to be a command officer (the more senior the better). After ICS, we shifted to requiring the first arriver, regardless of rank to become the initial IC. This change required us to train and prepare every responder to do some basic initial command stuff so they could establish command if they landed on the scene first. Having the first arriver become IC #1 gives us a fightin' chance that command will be set up in the front end of the event, and that a beginning level of command will be in place as subsequent responders arrive. The IC can now establish

Safety Effect:

and maintain an awareness of who's where, doing what. In most systems today, command is transferred to the first-arriving officer (from the non-officer IC #1). In the old system, everyone who arrived ahead of the officer (who eventually became the IC) typically free-lanced themselves into the hazard zone--if and when command got set up, IC #1 generally had no idea of what had happened (sometimes a lot) before their arrival. It was pretty much impossible to "get back" the control (opportunity) that was lost in those early (and generally very aggressive) self assignments.

It would be nice if we could always produce an experienced command officer, in a fully-equipped, snazzy vehicle specially designed and equipped as a CP, who always arrived ahead of the workers, and had command in place when the workers arrived. In the real world of actual fire department operations, a kid (wannabe firefighter) in a fire department delivery van, a fire inspector or two young firefighters in an ambulance sometimes arrive first. Simply, that person at that critical point is the only fire department representative on the scene. Based on this random arrival reality, the system must prepare, require, and support that person to evaluate the situation, give the IRR, deny entry to the hazard zone, and try to prevent harm until the cavalry arrives. The kid won't sound the same as someone who has been a battalion chief (BC) for twenty years, but they are all we have until a young captain or an ol' chief shows up.

Command Safety

Assumption, Confirmation, & Positioning of Command

Safety Effect:

Effective operational control = worker safety... effective operational control is the result of having an in-place IC and means that person can quickly initiate, manage, and change the position and function of hazard-zone workers. It's pretty easy to blab about the IC grabbing control of the position and function of the hazard-zone workers, and then maintaining that control as long as such a hazard zone is present. Being able to consistently do this requires first that an IC is set up and in place, and then that the IC is, in fact, actually doing the standard (eight) functions of command. Our troops live or die based on how (and if) this happens. Early and effective command/control will set the stage for how safe the operation will be from start to finish. When an empowered initial arriver becomes IC #1, we have taken advantage (both organizationally and tactically) of the very best up-front capability for that IC to take control of the position/function of whoever arrives on the scene--this is possible because the first person is there to "welcome" every subsequent arriver (i.e., everyone). This "welcome" involves having those arriving teams assigned to the IAP. The IAP must be consciously based on the overall offensive/defensive incident strategy, having their identity, assignment, and location entered into the logging/tracking/accountability system, and then specifically assigned by an IC to a specific organizational boss who looks after them.

Command Safety

Assumption, Confirmation, & Positioning of Command

Safety Effect:

Having the initial arriver take command is a very "quiet" process when it occurs according to the standard plan. What happens is streamlined, smooth, and almost transparent when viewed from the outside. Unfortunately, the confusion that emerges from not doing the standard command routine becomes very obvious when viewed from the inside and the outside. Having an IC set up at the beginning of the incident causes our individual action to occur within a consciously developed strategy and IAP where our collective effort and the effect of that effort is planned and integrated. What is not streamlined and smooth, typically occurs when the first arriver doesn't take command--now we start to develop a front-end, free-enterprise mess. With no one in command, arriving responders now assign themselves to "their own" incident. This uncommanded event now has a lot of aggressive firefighters all independently working (many times competitively) on their own self-selected operation. The scale and complexity of the uncommanded mess increases as more responders arrive and join the free-lancing free for all. It does not take long for this to begin to seriously affect the safety of all the hazard-zone workers who each have their own plan... this is how we give birth to a cluster.

Command Safety

Assumption, Confirmation, & Positioning of Command

IC Checklist:

☐ Use strong, quick, automatic command assumption to eliminate any command zero impact period (ZIP).

Safety Effect:

A major command system objective is to make the most accurate, correct, and safest initial decisions. When we can do that, the incident has a strong beginning and generally has a lot better/safer middle and ending. Not having anyone assume command at the very beginning of operations creates a period of free-enterprise action that is not under IC control--simply, such action can't be under IC control if no one has assumed and maintained the role of IC. During such ZIPs when there is no command in place, responders who arrive and put themselves to work are not assigned or accounted for by the IC (simply because there isn't an IC). ZIPs quickly become a command and control no-man's land. It is difficult (to say the least) for whoever comes into such an uncommanded (ZIP) situation and attempts to become the initial IC. This person must somehow convert the no-command ZIP (mess) from being out of any sort of command control to being under some kind of command control. Now the IC who is trying to gain control (once the ZIP has started) must do some "salvage command" (described in chapter seven) to somehow identify, "round up," and take control of the self-assigned free-lancers, develop an IAP, and then get these self-assigned workers into that plan. Many times, this requires some "Jack-the-Ripper"

Command Safety

Assumption, Confirmation, & Positioning of Command

Safety Effect:

type strong leadership and a forceful(!) command approach. In the ZIP operetta, the problem is not the workers (who are generally working hard to solve the incident problem). It's the person who was supposed to take command and didn't. When we miss taking command upon our arrival, we lose the streamlined, smooth, and transparent opportunity. "Jack the Ripper" must now take over to somehow gain control... "Jack" was not famous for being a smooth operator (but was well known for representing his very special management approach). Hazard-zone workers must be protected by strong, early, ongoing command that must be automatically assumed every time by IC #1 from the very beginning of operations. Strong, early, ongoing command ain't voodoo--it's having a real live IC start and do the eight real-live command functions from the real-live beginning. We know what they are (the command functions), what they look like, what happens when we do 'em, and what happens when we don't. The problem is that we can screw up the command process 1,000 times and the fire goes out (eventually) and nothing bad happens to us--on the 1,001 cluster, the note comes due and one of our guys/gals goes home in a black body bag. We must develop the understanding of the difference between being lucky and doing it right (luck makes us dumb). Every time we set up and operate, we had better be doin' the entire, standard command operational and safety routine because nobody knows when incident 1,001 is going to occur (and the current one many be it).

Command Safety

Assumption, Confirmation, & Positioning of Command

Safety Effect:

Free-lance fire fighting can be very effective for two basic reasons. The first is that it takes advantage of our most valuable asset: the spirit, skill, and motivation of our firefighters. They come on duty to kick the fire's ass. When we deliver them to the "OK Corral"--they come out with guns blazin'! The second reason that free-lancing can be so effective is that it is so quick. Simply, all the energy goes directly into the attack with very few distractions like following SOPs, having to pay attention to the IC, or blasting past basic safety rules. The problem with free-lancing is what we just blabbed about with the 1,000-times deal. Sooner or later, we will encounter situation number 1,001 that requires command and control of the overall incident. Then, if that level of management is not automatically in place, we're screwed. Twenty-five firefighters can (and will) quickly free-lance themselves into a hot zone, but they can't free-lance themselves out--an effective command system attempts to add safe to the positive part of free-lancing (energetic/quick) without wrecking what that spirited approach can produce. The addition of command and control requires we slow down a teeny bit, make more conscious decisions rather than emotional ones, and dedicate one person as the IC to perform the standard command functions. We need to be careful that we don't go to the "OK Corral" for our last gunfight.

Command Safety

Assumption, Confirmation, & Positioning of Command

Safety Effect:

Note:

Making the shift from free-lancing to incident command is a tough change. A major place where this struggle occurs is how IC #1 will behave at the very beginning of operations. The only way to create an effective level of command capability is to require that the initial IC always take command upon their arrival. Bosses must insure we do it every time. If the initial command assumption process (in the beginning) is slow, cumbersome or clumsy, effective bosses must invest the organizational support/energy to build the skills and confidence required to make it fast and smooth. The only way we can develop these skills and habits, both individually and organizationally, is by using the system (on every event) and then positively dealing with both unlearning old free-lance behaviors and learning new command/safety routines. This process becomes a critical part of improving the safety of the troops simply because bad habits can kill us and good habits can save us. Lots of times, we will stop doing what is accepted good practice because it is a new routine for us with a challenging (steep) learning curve. The system "smooths out" as we individually and collectively become more experienced, comfortable, and effective with the standard command process. Bosses must coach and encourage everyone through rough beginnings that are a natural part of the change process... old habits die hard.

Command Safety

Assumption, Confirmation, & Positioning of Command

IC Checklist:

☐ Report conditions and confirm command assumption with a standard initial radio report (IRR).

Standard IRR includes:

- clear alarm
- unit designation/on the scene
- building/area description
- obvious/problem conditions
- action taken
- declaration of strategy
- any immediate safety concerns
- accountability started (announce the initial accountability location)
- disposition of resources (hold/add/return)
- command confirmation with name.

Safety Effect:

The IRR requires the IC to focus on, evaluate incident conditions (from their position) and then assemble and organize their evaluation/plan so they can articulate the critical factors that are present from the very beginning. Our response system is typically going very fast, particularly when we first arrive on the scene. Everyone is anxious to go to work and to do their part in solving the incident problem. Based on this natural response momentum, we must build a rational, standard "pause" in

Safety Effect:

the IC. When the workers require resources, coordination or support, they simply want to commo with the current IC and get what they need, so they can safely continue/finish their jobs. We should take steps to ensure that every individual incident always has its own unique designation. There should not be two "Central Commands" even if there are two alarms on Central Avenue simultaneously. Change the second (third, etc.) event to a cross street or an occupancy name to avoid confusion. We must be careful that the simplicity and clarity of naming command does not cause a lot of what should be sector business on the tactical level to get "dumped" on the IC, who is operating on the strategic level, only because it's so easy to just call "command." When this happens, the IC must reinforce the organization by redirecting everyone to their appropriate places in the organization. Once the IC develops the incident organization by establishing sectors, then company officers should commo directly to their assigned sector boss. On the local level where the typical incident environment is fast and dirty, the basic, straight forward stuff works the best. Naming command and attaching that name (permanently) to the current IC is one of those beautifully simple operational and safety deals that is easy to do, understand, and use effectively.

Command Safety

Assumption, Confirmation, & Positioning of Command

IC Checklist:

☐ Select the proper command mode (investigative/fast action/command).

Safety Effect:

A major IMS objective is to create an orderly beginning for operations. This is particularly important because strong initial command becomes so connected to the safety and survival of our workers. The three standard command modes (investigative, fast action, command) create the capability for the initial-arriving company officer to use a standard system to match the initial command action they take to the conditions that are present. Selecting and using one of these three standard IC position options makes initial action predictable and quickly understandable. Basic safety is increased because everyone knows where IC #1 is and pretty much what they are doing (and where they are doing it), based on the standard mode that the IC has selected and announced. The three basic command modes relate to the initial positioning of company officers who are serving as the IC. These officers have the option of selecting an initial mobile position that combines action with command in situations where they feel their direct participation can significantly affect the incident problem, or the safety of their crew. Such officers must learn, understand, and practice the skill(s) required to effectively decide, based on incident conditions where (and when) it is functional and safe to be the mobile IC #1, and where it's best to stay put, and establish stationary command. Most systems

Command Safety

Assumption, Confirmation, & Positioning of Command

Safety Effect:

require command officers who arrive in a command vehicle (sedan, van, sport utility vehicle [SUV]) who are either the initial IC or the command transfer recipient to assume the stationary "in-the-vehicle" command mode upon their arrival. Major command and safety problems occur when a company officer starts command out in the fast-action mobile position, and then a subsequent-arriving command officer who is supposed to upgrade command by staying in the CP takes off on foot. Now we have two mobile ICs who are actively (and generally competitively) moving around the fireground (many times in opposite directions). Wherever we fail to upgrade a standard part of the IMS (i.e., go from mobile to fixed), we miss the opportunity to strengthen the system in a standard way. Many times this failure to upgrade also produces the front end of an unsafe cluster-based outcome. The officer who receives command in the command transfer process from a mobile IC must assume and maintain a stationary position inside a vehicle that is designated as the initial CP for that incident--this is a major part of what command upgrade actually means.

Command Safety

Assumption, Confirmation, & Positioning of Command

IC Checklist:

- ☐ Correctly position command to match and support the current command mode.

Safety Effect:

Standard initial IC #1 (company officer) position options:

- ☐ investigative (nothing showing):

  Mobile IC is on a portable radio, moving around--evaluating conditions/looking for the incident problem.

- ☐ fast action:

  Mobile IC is inside with their crew on a portable radio directly assigning initial arrivers with the IAP and supervising attack operations that are attempting to quickly solve (i.e., attack/overcome) the incident problem.

- ☐ command:

  IC establishes stationary, remote command in their vehicle from the beginning, assigning responders to IAP positions/functions.

IC #1 must pick and announce one of the standard options (there is no acceptable action/position in between or outside these modes). Particularly when the IC is in the fast-attack mode, they must report the position/function of their self-selected

Safety Effect:

assignment. The three positions create a standard status (position/function) for the initial IC and create a regular framework for where IC #1 will be, the "work" the IC will be doing, and the basic objective IC #1 will be working on that goes with that standard position. The standard positions create the capability within the team to reinforce mobile command (initially set up by IC #1) with a later-arriving stationary IC (who will transfer command from IC #1 to themselves and become IC #2).

The three standard modes give the company officer who lands in the IC #1 spot a standard set of options that are developed and agreed upon ahead of the incident, that connect initial command position and action to the conditions they find. The standard position options create a quick, simple way for everyone to know where the IC will be located, and basically what they will be doing in each basic operational/ command position. This approach creates an initial command routine that is predictable and dependable among the whole team.

Nothing showing/Investigate--means just that; we get a call... bells/smells, false alarm, honest mistake, etc. IC #1 arrives (on an operational rig: engine, ladder, squad), doesn't see any problem showing from the outside, gives an IRR, hops off the truck with crew intact, fully turned out. Everyone else stages. IC #1 goes inside to check out what's cookin' (sometimes literally). IC #1 is on a portable radio and is mobile.

Command Safety

Assumption, Confirmation, & Positioning of Command

Safety Effect:

Fast attack--visible working fires in house/ small commercial. IC #1 arrives on a fire truck (generally engine). They believe their direct (physical) participation in the attack will make a positive difference in the outcome (search and rescue, fire control, crew safety). They give IRR and quickly assign attack team (engine and truck) coming in behind them. Everyone else stages. IC #1 goes inside with portable radio supervising their crew in the attack. The mobile, fast-attack IC mode requires a lot of skill, experience, and a cool head. During fast-attack operations, IC #1 is in a schizophrenic position (combining command and action) until their command boss arrives, transfers, and upgrades command to a stationary, exterior, stationary CP position.

Command--these are big, fast moving, potentially hazardous, complex situations that require strong command from the very beginning. They are typically big deals that are very visible, and it is obvious the incident will not be solved by a fast attack. Their resolution will require a significant deployment (i.e., lots of responders) and will not be resolved quickly. Being able to start operations under a stationary IC, who is sitting inside a vehicle on a stronger mobile radio, using a TWS in a supportable position creates the command foundation that gets companies

Command Safety

Assumption, Confirmation, & Positioning of Command

Safety Effect:

(and sectors) started off in effective positions, performing the correct action that supports particularly the safety of firefighters in the hazard zone and the rest of the incident operation. Face-to-face command transfer can be made when a command boss arrives and many times (if IC #1's crew is large and experienced enough) the initial company officer IC will stay in command, move to a better command vehicle and the responder boss will become the support officer (SO). This approach creates a strong beginning to building a command team.

Note:

The three standard beginning routines for IC #1 not only describe where IC #1 will be, and what they will be doing, but the regular modes also describe where the IC will not be and what the IC will not be doing. As an example, there is not a standard mode for the IC #1 to be running in circles, in front of the fire building with their coat unbuttoned and no SCBA, screaming orders to anyone(?) who isn't smart enough to hide out on side "C." This becomes a big-deal safety factor in how the hazard zone is managed. If the situation can be stabilized with regular attack lines, IC #1 (company officer) can go fast attack, and simply make the incident problem (and the hazard zone) go away by being directly involved themselves (physically). If the event is bigger than what can be controlled with one or two attack lines, the first officer must stay in their rig and act like a strategic-level IC from the very beginning. Whenever a responder is injured/killed on the fireground, someone is

Safety Effect:

going to ask, "Where was the IC?" When we look back at the history of firefighter fatalities in the American fire service, the answer to that question is many times both sad and embarrassing.

The point of all this safety stuff is not to look back and be able to provide snappy answers to past fatality questions, but to look forward to eliminate the events that require such answers. Fire department bosses should always be able to explain to their member's parents, spouse, kids, friends, and attorney the standard practiced command operational and safety system that is in place, applied, and continually refined to protect their loved one, and to eliminate the awful sound of a body-bag zipper.

Command Safety

Assumption, Confirmation, & Positioning of Command

IC Checklist:

☐ Set up a standard CP as quickly as possible.

CP (advantages):

- stationary, in good (as possible) vantage point--now and in the future
- remote--outside hazard area
- place to sit
- inside vehicle--quiet, lights, protected from distractions, weather, darkness
- stronger communications/electronics capability
- place to build command team/for staff support.

Safety Effect:

A standard CP position creates the capability for the IC to effectively see, hear, evaluate, decide, order, and react from a quickly-established on-scene "office." Creating this standard CP position becomes a huge factor for the IC to quickly get in a position to "add up" what is going on all over (the "safety score") to protect the troops. The command system is organized around the task, tactical, and strategic level of operation. Each level must take the standard position to perform its regular functions. The system must realize that even though we many times begin command with a fast-attacking IC who is a company-level officer

Safety Effect:

(to do rapid rescue/put a quick hit on the fire), we simply cannot achieve a strategic level of command, until we have an IC in a strategic position.

Reasons for strategic CP positioning:

- ☑ best place to listen/talk/think
- ☑ best place to plan/react
- ☑ best place from which to deploy
- ☑ best place to view and review

The IC must establish that standard, strategic CP position to develop and maintain an effective, ongoing level of command and control. The IC must use the CP to initiate and maintain on-line operational control (i.e., assign specific companies to specific spots, with specific objectives, and a specific boss), to be continuously available to commo (first call/immediate answer), and to monitor and maintain an awareness of incident conditions and operational action, and to manage the changing welfare and survival needs of hazard-zone workers. Every incident does not require a big-deal, fully-staffed CP level of command to resolve the problem. In fact, most local incidents are handled by the initial-arriving engine officer taking command, assigning a residential/commercial response, and quickly solving the problem. When this occurs, the BC arrives, looks over what's going on, does whatever will help IC #1, and many times never says a "command" word. The strength of the system is how we automatically back ourselves up. This requires the response of a command-level officer whose

Command Safety

ASSUMPTION, CONFIRMATION, & POSITIONING OF COMMAND

Safety Effect:

IMS speciality is to come in behind initial-arriving (many times fast attacking) fire companies and set up strategic command, if the event is not quickly stabilized. We shut off the automatic escalation of the command system when the initial and early arrivers solve the incident problem--but the capability to make the system bigger is automatically on the road along with the initial-alarm responders and reinforcements. Having command officers coming in behind IC #1 who are trained, equipped, and prepared to upgrade the command and safety system creates our capability to make the management response match the operational and safety needs of the incident.

Note:

This CP stuff is really pretty simple. A long time ago, early humans figured it out... that's why we call them cave men (emphasis on cave). They discovered that it was pretty dumb to stand out in the freezing rain, in the dark, when they could go inside, light a fire, and eat a roasted yak chop (very tasty). Our (translated) cave is a modern Chevy Suburban with a heater, overhead light and Snickers bar. Modern ICs should remember how their wise old grandmothers would describe the not-too bright as someone "doesn't have enough sense to come in out of the rain"... duh!

Command Safety

Assumption, Confirmation, & Positioning of Command

IC Checklist:

- ☐ Begin to "package" command for ongoing operation and escalation.

Safety Effect:

How IC #1 sets up command at the very beginning of operations creates the foundation that sets the stage for the entire operation. Using strong early command, clear commo, and standard attack planning in the front end of the operation, insures that workers are protected by an in-place, lucid IC, who has effectively and safely started the IMS from the very beginning (no ZIP in the beginning). Early command must be in place so it can be reinforced and expanded, as required, to match the ongoing operational needs of the incident and the safety requirements of the firefighters. What happens (early command) in the very beginning of the event becomes the launching pad for this escalation. This standard, command-based beginning creates the capability for the IC to develop (i.e., "customize") a command response to match the size, speed, complexity, and safety needs of the event. When this command development falls behind the incident level (mid-point ZIP), it quickly becomes impossible for the IC to effectively protect the workers in the hazard zone--or simply, to be able to get those workers out of the hazard zone when conditions go from okay to bad (sometimes very quickly). What a command-based beginning (and middle

Safety Effect:

and end) really means is that the local IMS must develop, apply, and refine a system that routinely and effectively connects IC #1, IC #2, and then the development of a regular command team (IC/SO/senior advisor [SA]) that quickly develops the command and control capability to keep up with an active and expanding incident-power curve. This requires a practical, doable game plan where each successive IC builds on what the previous IC did, and a detailed local SOP on how command will be transferred.

Command Safety

Assumption, Confirmation, & Positioning of Command

IC Checklist:

☐ Correctly accept/continue/transfer command.

Safety Effect:

Using a standard routine for both establishing and transferring (i.e., upgrading) command, that is used by the arriving team, creates the role/function/relationship capability within the team for the IC to effectively and safely establish and continue command. Having a system that trains and prepares every officer to be ready, able (and hopefully willing) to serve as an IC generalist, and the personal ability to understand and operate in every basic IMS position, based on their arrival order, is the best organizational way to insure both a strong front end and for effective ongoing command and control. A major effect (and benefit) of developing this command and control versatility among the company and command response officers is creating and then always maintaining the safety of hazard-zone workers. Simply, whoever shows up is assigned by the IC to do whatever must be done at that point in the incident to create/support/expand the incident organization... and most importantly, they actually perform that assignment. Using a fast-attack company officer IC in the initial stages of an offensive incident, and then having a subsequent-arriving response boss transfer, strengthen, and continue command within the IAP developed by IC #1 from an upgraded stationary,

Command Safety

Assumption, Confirmation, & Positioning of Command

Safety Effect:

CP position outside the hazard zone (but with an effective view of the hazard zone) creates a strong, safe, under control, integrated local team approach, and becomes a major safety capability. Workers who go inside the hazard zone should have the confidence that the standard command process is automatically set up and in place, and operating (outside) to support their efforts and safety. When workers feel that the bosses have their IMS act together, they have a positive feeling that makes them safer and more effective. There is plenty for the troops to think about and pay attention to on the exciting task level without having to worry about all the grand exalted commanders bashing into each other, trying to get the best photo opportunity and sound bite. Creating this automatic command reinforcement becomes an excellent (actual) example of the current phrase, "I've got your back." Simply, workers should be safer because of their bosses, not in spite of them.

Command Safety

ASSUMPTION, CONFIRMATION, & POSITIONING OF COMMAND

This happens when command is assumed, confirmed, and positioned:

- ☐ A single IC is in place; IC command/control starts early (with IC #1).
- ☐ Command is confirmed and named--IC existence and identity becomes well known.
- ☐ Standard information is gathered, evaluated and translated (and transmitted) in the IRR--action then emerges from that information-based report.
- ☐ Initial action occurs within the IAP (yea!).
- ☐ There is an IC to commo with.
- ☐ Function number one creates a positive beginning (launching pad) for functions two through eight.
- ☐ IC assumes predictable, dependable (standard) command position based on conditions.
- ☐ IC has time (with level-one staging) to evaluate/process/decide.
- ☐ We start out in the correct strategy.
- ☐ It is a lot safer beginning.
- ☐ We do what we say we are going to do (in the SOPs), from the very beginning--event runs much better.
- ☐ Pay me now or pay me later... this is where we pay now (so we don't have to pay later).
- ☐ Life is good... bad days get better.

Command Safety

Assumption, Confirmation, & Positioning of Command

This happens when command is not assumed, confirmed, and positioned:

- ☐ No strong, initial, single command gets set up--simply, there is no IC #1--lots of confusion. } = ZIP
- ☐ Sometimes multiple, conflicting, roving, yelling ICs wander all over... troops hide out... fire burns up/building burns down.
- ☐ No initial incident report/information/confirmation/name.
- ☐ No IC #1 to commo with (i.e., IC identity crisis).
- ☐ Next IC gets run over by non-staged companies doing their own thing... free-lancing.
- ☐ Lots of task-level free-lancing which is costly/auto assigning leads to poor, unsafe, uncoordinated position/action.
- ☐ Firefighters may be in offensive positions under defensive conditions.
- ☐ Unsafe beginning produces> highly hazardous middle produces> sometimes fatal end.
- ☐ Nobody knows what's going on.
- ☐ Functions two through eight are at a huge disadvantage (doomed, unsafe future).
- ☐ Safety is just rumor.
- ☐ Pay me now or pay me later... this is where we pay later.
- ☐ Life is a mess--bad day gets worse for everyone... firefighters and customers.

Command Safety

Assumption, Confirmation, & Positioning of Command

COACHING VERSION

"When you become IC #1, confirm command assumption in the IRR, as soon as you arrive. Then act/sound like command on the radio, stay put in your engine/ladder/squad unless the direct fast-attack engagement of your body (physically) can make a difference in saving the kid, submerging the fire, or protecting your crew. Think before you transmit (breathe and focus)... try not to sound like an idiot on the radio. Give a lucid, calm, IRR that describes conditions and action. Set up standard command so there is something in place for your boss to build on (it's hard to transfer command if there aint' none to transfer). Try to safely load up as much standard front-end command and initial-operational action as possible. The early windows of opportunity offer the best intervention chance for local responders (us) to achieve offensive outcomes--fire customers with a lot of interior seniority are generally dead; and older, well-established, expanding fires are, or soon will be, defensive situations. Eliminate any ZIP or initial free-lancing by loading the beginning of the event with strong, empowered command--it's our best chance of starting under control, staying under control, and never losing control. The first-five minutes are worth the next-five hours."

Command Safety

Situation Evaluation

CHAPTER 2

Command Safety

Situation Evaluation

Command Safety

Situation Evaluation

Function 2

Situation Evaluation

Major Goal

To develop and use a regular approach to situation evaluation using the standard forms of information management to determine the presence and intensity of critical incident factors.

IC Checklist:

- ☐ Pay attention to dispatch information.
- ☐ Conduct a rapid, systematic, accurate size up.
- ☐ Use command positioning for visual information management.
- ☐ Use maps, preplans, and reference material.
- ☐ Record incident information on a standard TWS.
- ☐ Use companies and sectors as info, reporting, and recon agents.
- ☐ Use a standard information inventory to identify known and not-known critical factors.
- ☐ Determine the intensity and dynamics of critical factors.
- ☐ Quickly identify and react to safety "red flags."
- ☐ Structure and time info management around tactical priorities and firefighter safety.
- ☐ Evaluate current conditions and forecast future conditions along a standard scale.
- ☐ Continually reconsider conditions.
- ☐ Interrogate/interact with the owner, occupant, building engineer, technicians, etc.
- ☐ Maintain a realistic awareness of elapsed incident time.
- ☐ Consider fixed factors and manage variable factors.

Command Safety

Situation Evaluation

Author's Note:

As we developed the material in this chapter, it just kept piling up--it turned into a monster. What this reflects is how critical situation evaluation (risk analysis) is to the operational safety of our troops. It also shows how important this function is to what the IC does to protect the firefighters, and how critical the initial and ongoing size up is to effectively performing the other command functions.

We developed a set of "gauges" to describe the presence, intensity, and dynamics of the critical fireground factors, in what we hope is a simple and understandable way. There are more conditions, factors, and elements that the gauge approach could be used to describe. We had to stop (we think we got the biggies) because the chapter was turning into another book. Perhaps the reader can continue to apply the gauges to more stuff (please send your ideas along).

Thanks for your attention and patience as you read on.

Command Safety

Situation Evaluation

IC Checklist:

☐ Pay attention to dispatch information.

Safety Effect:

Responders should critically listen to the initial dispatch so they know the address, the basic incident problem, and the other companies/bosses who are assigned to that incident. This initial orientation creates a safe and sensible beginning to the response. Blasting off to the incident, not knowing these basic details, many times produces an exciting and screwed up (unsafe) "Keystone Cops" beginning. Previous knowledge that the commo center and the responders have of the people, places, and things at the incident site (response location) provides a very practical information head start about occupancy details, hazards, and the history of what kind of business (happy/sad, sane/insane, safe/hazardous, nice/nasty) we have done at that "address," with those people, and with all the regular and special stuff that has occurred during earlier responses to that place. Critical incident information should be gathered and given by the communications center and paid attention to by responders in the initial dispatch, particularly as it relates to safety issues. Clearly indicating the designation of the responders (i.e., unit/company numbers) is a critical piece of dispatch information. These individual units will become the basic work units that deliver the manual laborers, command system helpers, and the major safety focus (assignment/tracking/accountability) of

CITY OF PHOENIX, ARIZONA/FIRE DEPARTMENT
RESPONSE CARD
SD | EMS | FIRST ALARM | Time
2-1 | STILL | HAZ MAT | Card
ADD:
OCC:
① ② ③ ④ ⑤ ⑥ ⑦ ○
E
E
E
EP
L
L
S
R
U
BC
92-122D REV. 11/84

(For many years, PFD officers carried 3" x 5" incident response cards in the pocket of their uniform tee shirt. When they were dispatched on a call, they would routinely record the address, radio frequency and the other units that would be responding with them. The card also had a small space to record the assignments the officer would make to other units, if they became the IC. It was a very practical, low-tech way to capture and record the basic information that was required to effectively and safely manage an incident. This card evolved into what is now a tactical work sheet. The author witnessed a lot of early day ICs effectively managing fairly major incidents on the 3" x 5"... in fact, our tee shirts still have the pocket even though we now have on-board computers that display the same info that was on the old response card--now all us old guys use the pocket for our reading glasses.

Safety Effect:

whoever becomes the IC. Lots of operational and welfare problems occur when the IC tries to commo with units that are not on the assignment (but the IC thinks they are), or when the IC misses units who are actually on the call (but the IC doesn't know are actually on the scene). It sounds pretty simple for everyone to jot down who is going to the party, but just watch (and listen to) the mess that occurs when we don't get it straight.

To prevent deployment "free for alls," designated adults (company officers) must control their troops and respond only when they are dispatched... but in the real hometown world, clever, creative responders find ways to "take in" fun-sounding activity when they were left out. Units who self dispatch, come upon an active incident, or receive divine direction to respond must always follow regular level-one and level-two staging procedures, and must make their existence (arrival/location) known to the IC or staging officer (level two). This creates the capability for everyone who is on the scene (no matter how they got there... "legally" or "illegally") to be logged in, accounted for, and integrated into the regular accountability tracking and work cycle. In fact, units on the scene who were not dispatched can be in an extremely dangerous spot if they "sneak" in, secretly go to work, and then get in trouble--when this happens, the IC must get through the "where the hell did they come from?" natural reaction (i.e., surprise) that delays getting them help. IC bosses must subdue, provide strong direction, and inspire Ninja responders to follow the very basic deployment rule: don't go if you are not dispatched.

Command Safety

Situation Evaluation

IC Checklist:

☐ Conduct a rapid, systematic, accurate size up.

Safety Effect:

The IC must determine (evaluate/process/decide) the critical incident factors before an effective and safe action can begin. The initial size up must identify the what, where, and when of the incident problem. Size up means evaluating conditions, then thinking and deciding what to do about them. There is a big safety difference between rational action and emotional action (i.e., "emotional motion"). Investing a small amount of front-end time, evaluating and thinking about incident conditions, saves critical time later on, because it gets workers in the right (safe/standard) place, performing the correct action from the beginning of the event. This rational approach eliminates the need to "save" (i.e., rescue) workers who have jumped into dumb and dangerous places (i.e., "non-thinking action"). MOST OF THE TIME, THE FASTEST (AND CLEARLY THE SAFEST) WAY TO GET TO WHERE YOU ARE GOING, AND TO DO WHAT YOU WENT THERE TO DO, IS TO SLOW DOWN.

A major part of the size-up process involves evaluating the conditions that directly threaten our firefighters. These are, in most cases, the exact same hazards that also threaten our customers, who are in their pajamas, in contrast to us in our PPE. These conditions involve the structural integrity of the building, a profile of the thermal and toxic stage of the fire, and how the size, arrangement, and

Command Safety

Situation Evaluation

Safety Effect:

the complexity of the hazard area will affect our entry, our operational capability, and our exit. The IC must evaluate and compare these conditions against the basic protective capability of our safety system. The basic safety system is made up of an adequate number of trained, fit firefighters, PPE, operational hardware/water, safety SOPs, and IMS. A continual comparison of the safety system against the incident hazards becomes the very practical strategic (offensive/defensive) basis for deciding where firefighters can and cannot operate, and must always be a major function of the IC.

This hazard level/safety system capability comparison must conclude in a quick, fairly simple "GO/NO GO" response, in relation to selecting a potential operating position inside the hazard zone (i.e., offensive). The IC can be in the "GO" (inside) mode (offensive), where the situation evaluation indicates the basic safety system is adequate to protect the workers from the hazards that are present in that operational spot. Where the IC's evaluation indicates the hazards will out perform the safety system, the IC must create a NO-GO (inside) response in that hazard-zone location, unless (until) the IC can assemble, assign, and manage the resources that can increase the size of the safety system to match (or ideally exceed) the hazard profile. Lots of times, it's impossible to develop such a larger safety system (simply) before the building (or that fire area) burns down. In these cases, the IC must control defensive operations (i.e., "patrol the perimeter") that are located outside the hazard--collapse zone. Patrolling the perimeter simply means

Command Safety

Situation Evaluation

Safety Effect:

clearly identifying and announcing "defensive," either initially, or if offensive conditions change, creating an adequate incident organization (geographic/safety sectors) to cover the entire incident, create and maintain effective ongoing commo that keeps everyone connected to the defensive plan, and provide tough bosses who are assigned to positions and functions around the event who are not timid about directly and forcefully maintaining safe outside the collapse-zone positions, and quickly discouraging any stupid daredevil stuff.

Note:

Everyone involved in incident operations must develop a realistic understanding of how serious defensive conditions are to even fully-protected firefighters. Many times when we get trapped by such conditions, there is very little chance to be rescued. In these cases, the only effective response is to keep our troops separated from these defensive conditions, and prevent those firefighters from requiring rescue. We must guard against developing the unrealistic feeling that, if we have the entire safety system in place (rapid intervention, backup line crews, adequate sectors, etc.), we can go anywhere, and do anything. When the building falls down on us, when we run out of air, or when the fire poisons us, or burns us up, all the IMS and safety stuff on earth will not save us. The only effective response to defensive conditions is to not get them on you.

Command Safety

Situation Evaluation

IC Checklist:

☐ Use standard command positioning for visual information management.

Safety Effect:

Most of the time, the officer of the first-arriving engine company (simply because we generally have more engine companies than any other units) becomes IC #1. That person uses visual info during response (smoke, fire, explosions, falling meteors, etc.) to begin the size up and then, upon their arrival, looks directly at (and evaluates) conditions on the exterior of the structure/area from their initial position to start making operational, command, and safety decisions. If IC #1 (company officer) elects to take an inside, fast-attack position, they will then directly get more info on interior conditions, because they are (or soon will be) physically inside. What they saw on the outside, combined with what they find on the inside, provides IC #1 with a more complete picture of the status and effect of what's happening. When a command boss comes in behind the now attacking inside IC #1, and transfers command (becomes IC #2 on the outside), we achieve an even more balanced and stable ongoing in/out perspective (IC #1 who is a fast attacking mobile IC, now becomes "interior sector," or if the inside is large, one of the interior sectors/IC #2 is the command boss, who stays outside... we must always have

Safety Effect:

only one IC). We build the command team to support the IC, who monitors, supports, and directs the fireground organization and the action that is created through the regular IMS chain of command. There are critical factors that can only be seen or sensed from the interior, and some that can only be seen from the exterior. Being able to effectively have officers who are covering both positions, in a cooperative and coordinated way, begins to build and strengthen the incident organization, and will create a stronger inside/outside information balance. This balance will facilitate a safer and more effective operation. Effective, balanced, ongoing situation evaluation that receives, processes, and continually "adds up" and reacts to the collective information picture of the entire incident requires an outside CP be established as quickly as possible. In the CP, the IC becomes both the exterior "lookout" (from the IC's visual vantage point position... close to, but outside the hazard zone), and the person (IC), place (CP) process (IMS) who receives and reacts to all the reports on position, progress, and needs from all over the incident scene (particularly the inside/rear). The IC must connect these ongoing reports to continually create the best "picture" of actual conditions. How the overall "picture" looks (and changes) becomes the basis for the "safety math" that always is comparing

Command Safety

Situation Evaluation

Safety Effect:

the level of the hazard(s) present to the capability of the safety system. This comparison becomes the basis for how the IC manages the position and function of the hazard-zone workers. Everyone on the team uses the basic offensive/defensive decision and declaration as the common organizational term to describe the two basic attack and operational options (position/function) we use to protect our workers.

Note:

It's pretty easy to quickly say "the IC must manage the position and function of the hazard-zone workers"--it's pretty hard to actually do that. The IC must use the standard command functions as the way to actually pull off task/tactical-level position and function control. These command functions form the basic job description of the IC, and describe what the IC must do on the strategic level, along with the use of well-placed sectors on the tactical level that meet the actual needs of the situation to somehow manage where the firefighters are physically, and to influence where they are and what they are doing. Consistently doing all the command functions places the IC in the best position to do the most effective, ongoing, overall visual information management from a stationary, stable CP position.

Command Safety

Situation Evaluation

IC Checklist:

☐ Use maps, preplans, and reference material.

Safety Effect:

Maps provide the best routing, access, and water supply information. Preplans give the IC the key tactical features of the structure and the incident area. Preplans should emphasize the critical features that influence the safety and effectiveness of firefighters working in the hazard zone. Any other reference material will provide key information that the IC wouldn't otherwise have (or that could be difficult and time consuming to get... under typical incident conditions). This predetermined and prepackaged information puts the IC in a stronger evaluation, decision-making, and reactive position, leading to a safer operation. Written (hard copy) reference material used to support the command/operational process must be laid out in a simple, straight-forward format that makes it quick and easy to use. The information must be packaged and physically stored in a manner that is highly accessible, particularly in the beginning of the incident, when being able to effectively find and utilize information on critical factors can sometimes make a life or death difference. Being able to effectively use written material in the highly action-oriented environment (compressed time, high hazard, lousy initial info, etc.) of commanding a structural fire is a somewhat unnatural act-- such info must be developed in a realistic way, that is designed and suited to be used in the fast-and-dirty time and place where it will

Safety Effect:

be "looked at." An ideal approach to this challenge is to record (preprint) and store/do pre-fire plans on a TWS. This is the best of both worlds and provides a high impact, very practical information management head start, because the IC has the preplan recorded right on the sheet that will actually be used to manage the incident. In an earlier (hard copy only) time, we were trying to get both the incident reference information, and the way we stored and arranged the physical pieces of paper to somehow catch up to the command system. Today, that has almost flopped over--with computer-based data systems, global positioning system (GPS) oriented information formats, and electronic information systems, the potential information capability can quickly overwhelm the IC's physical and mental ability to survey, internalize, understand, and react to almost "too much" information. The trick is to package information/data up in increasing "bundles" that can be incrementally accessed, as the command system gets set up, and the command team gets into place. These "bundles" must get larger as the event goes on, and as the command team capability to deal with more information increases. The first-arriving IC generally can only handle a fairly limited amount of data, and needs simple, basic information on access, entry, and safety problems. The next IC is hopefully in a better "Suburban"

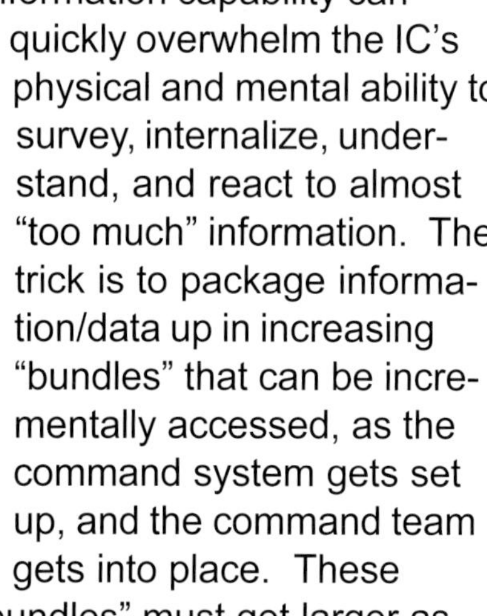

Safety Effect:

(i.e., SUV vehicle) oriented command position, and can deal with a larger amount (i.e., "bundle") of info. This information capacity increases as the command staff is expanded, and evolves to meet the needs of escalating situations. The ability to deal with more information must become a major part of the natural expansion of the strategic-level command team. Perhaps in the future (i.e., now), there will be a standard, structural fire command staff position of an "information officer" (not public information officer [PIO]) who shows up in the "info-mobile" and is hooked up electronically to a variety of tactical, geographical, architectural, structural engineering, process, storage, psychic, and human factors data base. That person could advise the IC and the command team (on line) about all the critical and related factors about the incident. How the safety system and the hazard level were relating to each other would always be the number one condition report that would be sent to the IC in the brave, new world of info management.

Command Safety

Situation Evaluation

IC Checklist:

☐ Record incident information on a standard TWS.

Safety Effect:

Response officers should develop the habit of writing down routine dispatch details, like the address, responding unit designations, radio frequency, and any other response details on a TWS. Recording this information should start at the very beginning of the response, and then be updated, as required, throughout the response.

Many times, units do not arrive at the scene in their usual order (surprise!), so whoever actually arrives first becomes IC #1. These non-normal arrival sequences can be the result of a lot of regular and special, different, many times very dynamic, local deployment conditions. When this happens, it can be surprising to everyone on the response--particularly the officer who is generally third due, who is the first arriver, and now must act as IC #1. It's a big help for IC #1 (no matter if they arrive in their regular order or not) to have a simple, usable, familiar form that has already been started, when all the responding officers got in their rigs and wrote down the address, the units on the assignment, and checked off the radio frequency. The form should have a standard place to record critical dispatch information, particularly

Safety Effect:

which units are coming in on the call. This makes it easier for the IC to begin to focus on who is responding, and then to be ready and able to assign those resources (as they arrive and stage) over the radio, and to record on the TWS who they are, where they are assigned, basically (briefly) what they are assigned to do, and when they were assigned (this starts elapsed time [ET] for managing self-contained breathing apparatus [SCBAs] and personal accountability reports [PARs]). When the IC encounters situations that require calling more responders, it's an easier and more natural act to add those units to a TWS that has already been started than to have to scramble to find a TWS and a #2 Ticonderoga pencil, remember and then attempt to recall, recreate, and record who was on the original call. The many times rushed beginning of an active and expanding event(!) is not a real neat time or place to begin a TWS. The IC's strategic role in the overall accountability system involves maintaining an awareness of who is where, doing what, working for whom, and for how long. The IC must use an up-to-date, dynamic TWS to maintain an inventory, tracking, and accounting of the units working in the hazard zone. Accountability locations must be recorded (just like sectors) on the TWS. The TWS is also the place where the IC records and keeps track of the resources that are both responding and on the scene, standing by (staging, rapid intervention crew [RIC], on-deck, resource sector, etc.), and available for assignment into the hazard zone. These are the troops who can cover new positions/

Command Safety

Situation Evaluation

Safety Effect:

functions, back up interior companies/ sectors, or go in to rescue the insiders (if necessary). The TWS is the standard (fast, simple, effective) form the IC uses which provides the basic info that protects the workers, so it is important that the sheet is started as a standard practice from the very beginning of our response, and maintained throughout the incident. An effective TWS is to the IC what a hose line is to an engine company--it's our on-line location/status basic information life line, when everyone must react quickly, because what's not supposed to happen, happens (i.e., lost firefighter, collapse, etc.).

Command Safety

Situation Evaluation

IC Checklist:

☐ Use companies and sectors as info, reporting, and recon agents.

Safety Effect:

A major information management challenge for the IC is to quickly develop an incident organization that covers the critical (i.e., hazardous) areas of the scene. The incident organization is created to both conduct operations, and to report on conditions in their area/function, while they are doing their assigned job. Using the incident organization to routinely build this operational and reporting system creates the capability for the IC to receive, evaluate, process, and react to current conditions (particularly what the IC cannot see from the CP) and actions from all over the incident site. The IC must use this same info management system to integrate that decentralized information to develop a forecast for the future for the entire incident. A lot of bad things happen to workers because hazardous conditions are present, and the IC simply does not have (or use) the regular, decentralized, organizational capability to find out about them, and to report on those conditions in time to do what it takes to protect the workers. Building and using a two-way commo system that can quickly determine (based on their position) and exchange critical information becomes a huge part of the overall safety system. This system creates the capability for the entire

Command Safety

Situation Evaluation

Safety Effect:

this yes

response team to effectively connect what they are doing (and seeing) to actual, current conditions, and to maintain the operational agility to always "stay ahead" of changing conditions. Not having this critical (unknown safety) information is a big deal, and it has cost the lives of many firefighters. Having operating units transmit ongoing reports of position/progress/needs, along with making ongoing condition reports, as a standard part of every assignment, creates the regular, natural, habitual capability for the IC to quickly develop the information that is necessary to manage and move the troops, if necessary. This decentralized company/sector based reporting approach creates the capability for the IC to stay in the CP, to be continuously available to receive information from all over the scene, and continually connect, integrate, and interpret how conditions affect worker safety. What this means (simply) is that the IC assigns sector officers to the inside, top, and rear (as an example). These sector officers directly supervise fire companies right where they are assigned and working. Sectors become the eyes, ears, and nose of the IC all over the incident site. The IC then is able to stay in the CP and exchange info with these off-site (from the CP) command partners. If the IC needs info from that sector position, they contact the sector officer and ask for ("info pull") that piece of info. Company/sector officers automatically and routinely report ("info push") anything

Safety Effect:

they see in their area/function that is critical to the IC. Using the incident organization, as regular reporting agents, eliminates the need for the IC to continuously physically "lap" the building (pant!) to keep themselves informed. In order for the system to work, everyone must understand their role, the standard position/function that goes along with that assignment, and they must trust each other (big deal). Such trust is the result of a strong, well-practiced system, positive relationships among the team, and the organizational capability to translate ongoing operational experience into continual system improvement.

this no

Effectively using sectors and companies to report conditions from their assigned position/function requires a special (different) set of personal IC skills and a different basic organizational approach. This is particularly true if that IC has previously evaluated conditions by physically going and directly viewing, sensing (i.e., feeling), and smelling those conditions in an "earlier life." Receiving info from a recon source, rather than being physically and personally close to that condition, is a fairly normal challenge that occurs in making the transition from a company to a command officer. Being physically close to things (particularly exciting things), as they are occurring, naturally "feels good" to action-oriented people (like us), and it is somewhat unnatural to have someone else report these conditions to you

Safety Effect:

(mostly over the radio). The decentralized company/sector/IC information exchange requires the IC to discipline themselves to stay put in the CP, make assignments to critical places, and then receive (listen!) and react (pay attention!) to what the officers in those places report. The alternative is for the IC to hop out of the CP, do some aerobic roadwork ("orbiting"), and to almost instantly lose control of being able to effectively and safely assign, and then manage the position and function of the troops operating in the hazard zone. The basic IMS safety system calls upon the IC to stay put inside a CP vehicle, primarily so they can do their strategic part in protecting the hot-zone workers. The most absolutely critical, potentially fatal report the IC will ever receive from any hazard-zone position is when some firefighter faintly gasps, "Mayday," over the tactical radio channel. This is show time for the IC and the entire fireground command and safety system. What the IC has set up and what is already in place (correct strategy, adequate resources, strong command, good sectorization, tactical reserve, etc.) will now determine if there is going to be an effective rescue, or if three days from now we are going to get to listen to some very sad bagpipe music. If the IC has not automatically done the entire, regular, agreed upon (i.e., standard) command and safety routine on every one of the 500 incidents ahead of this one, it is too late for Firefighter Smith, who now is lost, three and a half lengths of hose inside a burning building, with a yelping pass unit, and a SCBA

Safety Effect:

low-air alarm going off... simply, it's impossible to repack your parachute halfway down.

The IC must always "managerially" command, control, and protect our humans, as they attempt to "operationally" stabilize/reduce/eliminate the incident problem. When that problem is (in fact) offensively controllable, the IC moves the workers closer to it; when the problem is not directly controllable (i.e., "defensive"), the IC must keep or move the troops away from it, and to control their position outside the collapse zone. We use the basic offensive/defensive approach to simplify this, so that we can quickly do standard, safe, positional management under difficult incident conditions. The IC cannot consistently do this successfully at Mayday time, unless the entire team has practiced and refined both ends of the decentralized organizational plan. A major part of that two-way plan requires the IC to capture (and practice... a lot) the standard, complete incident command routine, that is absolutely required for them to first control their own (the IC's) position and function, so they can always be in a position to effectively control the position and function of the hazard-zone workers.

Command Safety

Situation Evaluation

IC Checklist:

- ☐ Use a standard information inventory to identify known and not-known factors.

Safety Effect:

The IC must develop an understanding of the standard inventory of critical incident factors which can injure or kill firefighters, that can be present at structural fires. The IC must approach the incident as if all these critical factors are present. This knowledge creates the capability to use this standard mental inventory, as a checklist, to sort out and address what is and what is not known about these critical safety factors. The critical structural fire fighting factors generally revolve around a combination of basic conditions, that do the following, very-bad things to our workers:

- ☐ The structure collapses on us (or under us) and crushes/traumatizes our bodies.
- ☐ The building and contents physically restrain or trap us, prevent our exit, and then we run out of air.
- ☐ The interior arrangement confuses us; we become disoriented, lost, and then we run out of air.
- ☐ The superheated and toxic products of combustion choke and/or burn us.

The IC must apply a quick, practical info management system (based on the critical

Command Safety

Situation Evaluation

Safety Effect:

factor inventory) that provides accurate intelligence, on what is known, and what is not known, about these present/forecasted conditions that cause these critical, standard, and non-mysterious, short-term (many times almost instant) fatal hazard factors.

The IC must evaluate structural/fire conditions, accountability, maintaining a viable (protected) ongoing exit for inside workers, and an accurate evaluation of fire/thermal/toxic conditions. The possible consequences of these fatal hazard factors are so potentially severe, that the IC must not expose workers to situations where the extent of those hazards may be serious or severe, until adequate info is available on the degree, location, and details of that hazard--many times the potential, but in some cases, unknown presence or severity of these unknowns become the basis of defensive operations. Simply, the IC must keep the troops away from these unknown (but potentially present) hazards because of their severe consequences, until an effective level of information is gathered and processed.

A big-time IC situation evaluation function involves sorting out what is and what is not known about critical factors... and then going to work on what is known, and using unknowns as information targets. The IC must then make assignments to the locations and functions that can evaluate and report on those targeted pieces of unknown

Command Safety

Situation Evaluation

Safety Effect:

information. This reporting function becomes a major reason why the IC must quickly develop an incident organization, that can "cover" the critical incident locations/functions, where they can perform both operational action and provide evaluational reporting of conditions in that position or function. We put a strategic-level IC in place, as quickly as possible, to do the standard functions of command. The IC becomes the nucleus for all critical factor information processing. Initial incident operations begin with the IC conducting those operations, based on assumptions and their best guess (experience) concerning the critical unknown factors. The IC expands their info gathering capability by building an effective organization in key operational positions. A major safety target is achieved when the IC gets a set of safety eyes (sector officer) in place, in all of these key positions. This system reaches full activation when these positions report back to the IC and confirm/verify/or refute the size up of these critical factors. These position reports must verify the conditions (products of combustion), building design, and features (example: building appears to be a single story from the front-side CP view but, in fact, it is two stories with a basement in the rear). A practical knowledge of the standard inventory of factors, that come with that type incident place and problem, becomes the basis of the IC actually understanding this process. Simply, the IC can't know what is still on the unknown list,

Command Safety

Situation Evaluation

Safety Effect:

without a knowledge of everything that is included in the complete inventory of standard and special factors that go with that situation. Unknown, but present, critical factors have surprised lots of ICs, and injured and killed lots of hazard-zone workers. The IC must always be aware of the critical factors ("red flags") that lead up to (the "front end") and cause the potentially fatal conditions to slowly or suddenly actually become fatal. Being able to sort out the incident inventory into knowns and unknowns is a big-deal IC information management function. The capability to quickly and effectively deal with critical factors is produced by reflection and road rash (i.e., experience), that emerges from study, actual application, and review... and then storing the war stories and their lessons in the "memory banks" of the local IMS and the IC's noodle. This is why there is no substitute for the experience (wins, near hits, and hits) and the instincts that emerge, simply from responding to and living through a lot of incident operations. We must both educate, and then trust, our "gut instincts." If it feels bad, there is probably something going on at a very subtle level that is sending a quiet message. Paying attention to such a cue has saved a lot of us, when some unplanned (surprising) event quickly occurred in a place or thing that we moved or stayed away from, but didn't know exactly why. Sometimes, the hazards shout at us, and sometimes they whisper--being tone deaf can be very dangerous.

Command Safety

Situation Evaluation

IC Checklist:

☐ Determine the severity and dynamics of critical factors.

Safety Effect:

A major incident evaluation function involves the IC using the standard critical factor inventory, as the basis to identify the critical factors that are present, and to then determine the severity of that hazard. The IC must use visual information from the CP, and reported information from the incident organization (that must be assigned and in place all over the incident site), to make this critical factor determination. Critical factor severity is a big deal, because these are the conditions the IC must consider to develop an IAP that will convert the incident from out of control, to under control. The severity of the factors will determine the action that must be taken--the more severe the factor, the more severe the action must be to solve the problem. The IC can only match action to the current and forecasted conditions by escalating/repositioning fire fighting and rescue operations throughout the course of the incident. If conditions continue to worsen beyond the effective and safe level of offensive operations, the IC must quickly move the troops to safe, defensive positions, and then "surround and drown" from exterior positions, that are outside and away from the collapse zones. These factors are also highly dynamic and will become more or less serious, as the incident goes on, and as action is

Safety Effect:

applied. Based on these changes, the IC must track (and record) if those factors are getting better or worse, and the time frame of any changes that are occurring or are about to occur. In order to effectively exchange this information, the team must decide (and agree on ahead of the event) what the standard stages (levels) of incident conditions look like, what is the relative hazard of the various levels, and what is an effective (safe) response to each level. For the critical-factor information management system to work, the decentralized reporting units (companies/sectors) must report the existence of those critical conditions, and must also continually update the IC with reports on how the intensity of those standard conditions is changing. This requires we develop and use a standard, descriptive "language" that connects everyone. Major problems occur when the troops on the inside, and the IC on the outside, don't agree on the presence of critical factors, or on the level/intensity of those factors, if they are (obviously) present. This info management process also requires a quick, accurate way to describe how those factors are changing, and the correct action that is required to deal with those conditions. That language must be short, quick, and easy to understand/exchange.

Situation Evaluation

Safety Effect:

We will use a set of gauges to describe and package information about the existence, intensity, and changes of the major conditions the IC must consider throughout the incident. The gauges provide a standard way for the IC to "hook up" the various critical factors to a system, that graphically shows (using a standard visual format) what's going on. The IC can then use the gauges to "read" the situation. The gauges create a format that is familiar, visual, simple, and quick to read and understand. They describe the relative intensity of that factor, and we can use the same format to show a variety of different conditions. Gauges have no ego, rank, or politics (when the gas gauge is close to "E," simply, you're about out... duh). These incident information gauges only provide visual information about whatever condition they are "hooked up" to. They do not create any automatic action, so they require a human to "read the gauge," and then create a response to the information the gauge is providing. In the case of our safety system, the IC is the person who is in the best position to receive information from all over the incident, initially set the gauge, and then keep track of its current setting, by continually "resetting" the gauge to reflect changes in that critical factor, based on visual info from the CP, and reported info coming into the CP from all over the incident organization. To be an effective gauge manager, the IC must have a basic understanding of the dynamics and details of the conditions the gauges are showing. Simply, the IC is the person "driving" the

Command Safety

Situation Evaluation

Safety Effect:

incident and is in the best position to see the "dashboard" (with the gauges)... some smart person called this process "dashboard metrics."

Based on the need to be brief, simple, and easy to understand, hazard levels on the gauges are expressed on a one (1) to five (5) severity scale. The scale and the standard reaction to the levels are applied to all the critical incident factors that involve hazards to firefighters.

		hazard severity
green	okay...	1. discernible--conditions are essentially okay
yellow	present...	2. clearly present and beginning to create an operational hazard
	serious...	3. creating a serious hazard
red	severe...	4. widespread and extremely severe
	fatal...	5. deep seated and fatal

We must always apply the standard safety routine to every level.

Our standard reaction to hazard severity:

1. We can move around the inside, investigate, and take action.
2. We can extend a well-protected fast attack.
3. We can operate inside, if conditions are improving/if conditions are worsening, we must begin to exit the hazard zone.
4. We can no longer conduct inside operations.

Command Safety

Situation Evaluation

Safety Effect:

5. We must stay out and away from these inside conditions.

Note:

The gauges attempt to create a real simple, conceptual method to package up critical factors. They are not highly scientific (to say the least) and are only meant to describe a range of conditions that go from okay to fatal. Please don't get hung up, as you go through them, on the minute details--they are designed to help us create a basic way to define, understand, describe, exchange information, and react to fast-and-dirty incident conditions. We must many times operate in a pretty rough place, and it helps if the team has had a chance to create and use a system, that standardizes how we approach and manage the conditions that can hurt/kill us.

Command Safety

Situation Evaluation

OVERALL INCIDENT RISK LEVEL:

The overall incident risk level gauge is the "master gauge" for the entire incident. This gauge must be set to reflect the total of all the other individual gauges that are each connected to their own critical incident factors. Our safety and operational approach is based on a standard risk management plan. The plan creates a connection between the risk we will take to the benefit of that action--big risk to protect savable lives, little well-managed risk to protect savable property, no risk at all for lives and stuff that are lost. While the risk management plan is short and simple, it is only effective to the extent the IC can accurately determine the overall risk level. Many times making this determination can be very challenging, and requires the IC quickly "hooking up" the gauges to the critical factors to determine the initial, overall risk level... simply, it's impossible for the IC to accurately select the correct risk level, if they can't set this gauge. This initial setting then becomes the basis for choosing the beginning strategy, and for managing the tactical

Overall Incident Risk Level Condition

1. low
2. low +
3. medium
4. medium +
5. high

Overall Incident Risk Level (contd)

priorities (within the IAP). No matter what level of risk we take, we always conduct operations in a highly calculated manner that reflects the consistent application of our basic safety plan: an adequate number of trained, fit, informed, clear thinking, rational firefighters/PPE/hardware/water/safety SOPs/IMS. A major IC function is to initially set the master gauge, and then to continually keep it adjusted by looking at incident conditions, and listening to, and exchanging information from all over the incident, to use all the other gauges to produce the ongoing overall situation evaluation. The IC must always align (and adjust as necessary) the strategy and IAP to match the master gauge reading. This process is pretty simple, quick, and real important, and like most things that are pretty simple and real important, it takes lots of brains, guts, experience, and team work to make it work. If you sign up to be an IC, you had better sign up to do the personal art and the professional science required to manage the master gauge. To do this effectively, you better be prepared to do equal parts of hockey, chess, and ballet all at once.

Command Safety

Situation Evaluation

BUILDING SIZE/AREA

1. Small

Condition

1. small
2. medium
3. large
4. huge
5. ultra

A major safety factor involves the size of the area where firefighters are working. The potential hazard level is directly connected to the size of the hazard zone--the bigger the area, the bigger the potential problem. Another cheerful reality is that larger areas simply create more places for firefighters to get into trouble. Big areas simply create the need (and the possibility) for us to go farther into the hazard zone. The farther we go, the more dangerous it becomes, and the more challenging it is to provide protection to the crews who are operating in those positions. Managing these operations can be very difficult. The IC and the entire command and operational team must effectively adjust operations to match the size of the incident area. This requires (as the area increases) that we "ramp up" all of our command, control, and safety systems in a conscious, deliberate, and practiced way. We get our asses kicked when we take 1,400 square-foot house fire tactics (and mentality) into a 50,000 square-foot warehouse. Most modern, progressive fire codes require sprinkler protection for areas starting at about 5,000 square feet. The folks who insure these

BUILDING SIZE/AREA

2. Medium

BUILDING SIZE/AREA

3. Large

Building Size/Area (contd)

places figured out that normal, manual, fire-suppression capability becomes seriously challenged at about that size--unfortunately, about a bazillion of these bad boys were built before such sprinkler protection was required. Now they are sitting on every street corner in the older sections of all our towns, just waiting to create the offensive looking conditions that seduce two engines and one ladder to go inside (clear to the back), and then sucker punch them with a big blast of flashed over hot stuff that really complicates their exit, and provides a huge surprise to the half asleep IC's understated, undercapitalized IAP. Wily, clever ICs don't underestimate any tactical situation, from tiny to huge. No matter their size, they all quickly produce the same deadly elements (bigger ones just produce more of it). The IC must quickly focus on creating an operational response to match the size of the incident area. The IMS is highly modular, so when we go into a large-area building we can expand the operational and command system to do stuff like doing a real accurate hazard analysis, quickly calling more alarms, increasing the size of the accountability system, assigning more safety officers, expanding the incident organization including overorganizing (i.e., carefully managing) interior sector command and control, and doing early and ongoing hazard analysis, and all the other organizational stuff that guarantees a round trip for all the firefighters.

The IC must always connect the actual operational capability of the resources that are on the scene to the size of the place where they must operate. In some cases, this requires

Building Size/Area (contd)

the IC to order companies to not operate in large areas until enough resources are assembled and effectively organized to safely go into those big places. Sometimes what this means is that the IC must either set up offensive operations on the outside by laying lines, placing ladders, forcing entry, or just manage these operations in the defensive mode until there are enough resources on the scene to safely operate offensively. In these situations (big hazard zone/not enough resource), the IC must resist the temptation to just "do something"... particularly, if that "something" is to put firefighters in an interior position, where they can get in, but may not be able to get out. The IC must pick the time and place to fight, and then take enough resources with them to get the job done. Sometimes there is enough in the beginning, and in these cases the IC can do a quick attack. Other times, there are not enough people and stuff to safely attack, so the IC must "procrastinate." This requires the IC to control the troops and do whatever can be done to get ready for the calvary to arrive. Smart, safe ICs learn to pace themselves and to always develop an overall strategy and an IAP that connects resources to incident conditions. This reality becomes another compelling reason why such places should be protected by automatic, built-in suppression (sprinkler) systems... it also must create the realization that even the most well-executed manual fire suppression, applied in the middle of the fire, cannot be made as effective as regular automatic sprinkler protection, activated at the very beginning of the fire... simply, the

Building Size/Area (contd)

most capable firefighters on the planet (great people) with hose lines, who arrive in the middle of the fire (very dangerous time), are not a substitute for a sprinkler head hooked up directly to a eight-inch supply pipe, that is waiting right over the fire before it really gets going. This is not a criticism of the firefighters. It is only a very simple description of reality... when a fire occurs that requires manual fire suppression, it is pretty much the result of a failure someplace in the fire protection system. Somehow the design, maintenance, code enforcement, and management has screwed up, and now E-1 must put on their fire-resistive smurf suits and somehow deal with an active fire. It's a big deal for everyone on the team to understand (and then act on) what it actually takes to produce a round trip, and the condition that can wreck that trip.

Command Safety

Situation Evaluation

FIRE STAGE:

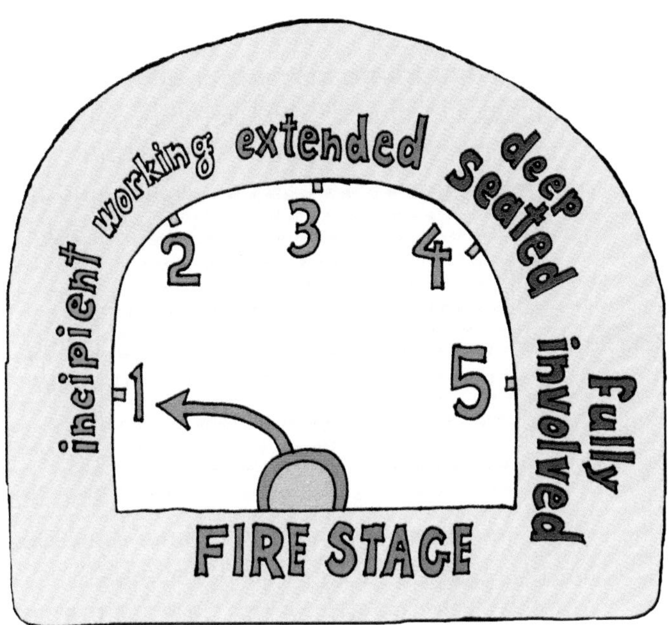

Fires in buildings burn through a series of fairly standard stages. The IC must evaluate the current and forecasted stage, and then connect the appropriate fire fighting action to that stage. This can be difficult to evaluate because fires are just naturally sneaky. They do stuff like hide in concealed spaces, burn in the back when the IC is in the front, and produce a lot of smoke to hide what's really going on. The IC must organize the incident and begin to receive information from all over to determine the fire size, location, direction, and avenue of travel, and what's left to burn. The IC must determine if the fire involves contents, interior finish and/or the structure. Adding all this information up produces a perspective on the percent of the fire area that is involved. As the percent increases, we quickly lose the effectiveness and safety of interior operations, so the strategy must shift from offensive to defensive. Making that strategic call is a major IC safety function.

Fire Stage Condition

	Stage	Involves
1.	incipient	contents
2.	working	contents
3.	extended	tweener
4.	deep seated	structure
5.	fully involved	structure

Situation Evaluation

Fire Stage (contd)

Lots of firefighter casualties have occurred in situations where inside operations were emotionally initiated ("candle moth" syndrome) before an accurate and rational determination was made on the size and location of the fire, and then the fire surprised the firefighters in a way that trapped them. We must not be distracted by the simplicity of structuring operations around the basic offensive or defensive strategic decision--the foundation of that critical decision revolves around the stage of the fire. The fire stage is directly connected to how long the fire has burned, so a major IC evaluation function is to determine the "age" of the fire. It is a major mistake for the IC to operate at a well-established, deep-seated ("middle-aged") fire, as though it was a situation that was in the beginning stage ("youngster"). Being able to get this gauge and keep it effectively adjusted becomes a major safety factor because as the fire becomes more severe, so does the hazard.

Command Safety

Situation Evaluation

PENETRATION INTO HAZARD ZONE:

A major safety challenge that occurs in conducting offensive operations, in a large commercial building, involves the attack distances (i.e., penetration) that are required, many times to simply reach the fire. A potentially serious safety problem occurs when we just add more hose to our normal attack lines, when we encounter tactical situations with long-attack distances to gain such reaches, and then attempt to operate with no other increase in our operational approach (resources, command level, exit hardening, sectors, increased accountability, more rapid intervention, etc.), as if we were twenty-five feet inside a 1,200 square-foot single family residence. Fire prevention codes require a standard (i.e., protected) corridor in places where occupants must go significant distances to reach an exit. We must develop a "corridor mentality" in how we protect fire fighting exit ways in those same places. This requires us to provide hose lines, lights, RIC teams, sector/safety officers, an increased level of accountability and usher/escort/guide crews that are assigned, operate (and stay) along hose lines, and other

Penetration into Hazard Zone Condition

1. normal distance
2. small stretch
3. big stretch
4. too damn far
5. fatally far

Penetration into Hazard Zone (contd)

interior access ways being used by crews, to protect and "harden" such exit ways. We must develop such a standard (i.e., reinforced) response to large-area operations, that requires the IC to quickly identify these larger areas/longer distances, and then immediately call for the resources that are required to effectively set up adequate protection for inside operations. Most departments have not historically developed SOPs, practiced or refined such long-distance, offensive, attack evolutions. These situations will continue to be very dangerous, until we develop, practice, and apply this approach.

Because we do operate at commercial building fires that are well involved (either upon or after our arrival) beyond the offensive stage, most of us have developed a defensive commercial command and operational plan. These defensive incidents basically involve an exterior "surround and drown" approach. We use this defensive plan when the fire has reached the point where it is obviously (or instinctively) impossible to go in/stay in. In these well-involved commercial situations, the fire makes the strategic decision for us, and we just go along with that decision and set up and operate on the outside. On the other end of the strategic scale, most fire departments do not have a refined, practiced, commercial building offensive fire fighting plan. When fire and structural conditions in large commercial places will allow entry and interior operations, the situation "invites us in," and a lot of times we typically apply our house-fire attack

Penetration into Hazard Zone (contd)

process, simply because we mostly go inside houses that are burning. The safety/survival problem is that we are now in a place that is, in many cases, fifty times bigger(!) than a single family residence. Such commercial places quickly out perform our regular, day-to-day operational systems. We can get into big trouble, unless the IC can quickly identify and react (i.e., deploy) to match those commercial conditions. Commercial building incidents that are in early stages can be particularly dangerous, because there is a visible problem present, entry (at that point) is easy, and we just naturally go inside to find and solve the problem (why we responded). Huge safety problems then can occur when conditions quickly change (worsen), and crews are a long way inside--now, they must either eliminate the incident problem (sometimes very difficult), or somehow exit back through the products of combustion that have accumulated behind them, and now those fire products make that exit journey very exciting, and generally very dangerous. Lots of times these crews (initially) were in an "investigative" mode and had little or no tactical support (water, tools, workers), and no "hardening" of the exit ways and exits. The IC must quickly read the attack distance/penetration gauge to develop an appropriate deployment and support response (order early/order big) to protect those firefighters who must go a long way into the fire area to reach the incident problem.

Penetration into Hazard Zone (contd)

Note:

Most of our training and drilling on commercial buildings involves outside, open-air work. We typically get together with our BC and two to three neighboring companies, and have a Sunday morning drill. Sunday morning is a good time because the schedule is generally open. Not much traffic is around, and we can schedule the drill before our big Sunday brunch/lunch. We pick out a familiar location that has lots of room (and good drainage). We hook up to hydrants, lay supply lines, put up ladder pipes, set up deck guns, and have an exterior H_2O movement/management festival. This is good stuff--it prepares us to effectively do the heavy-duty hydraulic (surround and drown), open-air, outside operations that are required to contain and control large defensive fires. The problem is that most of the time, when we do these defensive drills, we simply don't drill inside commercial buildings. Inside drills provide practice (and big-time reality therapy) on what it actually takes to effectively and safety operate in large areas, where we must support longer attack distances. These drills inside commercial buildings require that we greatly expand the residential attack routines that we apply a lot, get comfortable with, and get good at, to a larger, more complicated, and dangerous place. We will never develop the skill and ability to do safe and effective inside (offensive) fire fighting in commercial buildings, until and unless we give responders the opportunity to train on how to escalate

Penetration into Hazard Zone (contd)

all of our "everyday" (i.e., mostly house fire) operational command and safety systems to match bigger places. This ongoing practice and drill experience will result in a practical (and many times sobering) understanding of the capability and limitations of these expanded systems, and how long it takes to actually get them in place. This training will also produce a realistic awareness of where we can safely go, and what we can do in bigger, commercial areas/situations. In most cases, this training will require us to acquire vacant commercial buildings, as drill sites, to conduct this inside training. These buildings must be equipped with adequate training props and simulated interior contents to create a realistic place to apply and practice our operational and command systems.

Heat Condition

1. 200°
2. 400°
3. 600°
4. 800°
5. 1000°

HEAT:

Heat is generally one of the major problems the firefighters are trying to make go away. When the fire goes out, everything gets better. We use water to move the heat away from the inside to the outside, so we are effective to the extent we can make this hydraulic transfer. The challenge is that, currently, the only way we can do this in offensive situations is for firefighters to go inside buildings that are on fire, and apply water directly with hand lines. This requires those inside firefighters to get very close to the fire. We protect the hand-line troops with fire resistive suits, designed mostly for interior structural fire fighting operations, along with SCBAs. Modern technology has created big advances in the effectiveness of this PPE, to the extent that the gear will hold up to a flash-over for a brief period. The problem is that

Heat (contd)

we didn't change the basic anatomy and physiology of the firefighter we put inside the suits. At about 125° F the firefighter gets really uncomfortable, and about 150° F the human brain sends the central nervous system the following message, "What the ____ am I doing in here?" Let's say a flashover is about 1000° F. We can quickly see that there is about an 800° F design flaw they made at the original Ajax human assembly plant, in how the human body (in comparison to our turnouts) reacts to heat. Everyone operating on the fireground must learn to read the signs and symptoms of heat being generated, and then develop attack operations that realistically consider (and connect to) the heat profile of the incident (where and how much). The entire team must also routinely exchange information about the heat profile in their assigned position. The IC must continually "set and adjust" the heat gauge to monitor and manage the position and function of the troops working in the hazard zone (sometimes also called the "hot zone"), and must never lose sight of why we call it the "hot zone." Essentially, the IC is playing a heat management game, where we are trying to put water "on top of" the heat. It is useful for the IC to draw a simple diagram of the incident, mark where the heat is with a red marker, and then mark where the water is being applied in blue. Fairly quickly, whichever color is the biggest will describe the basic outcome. During this process, the IC must protect the firefighters from the red, as they apply the blue... the IC must keep a

Heat (contd)

careful eye on the heat gauge as this goes on. We can assume that nothing is completely fire resistive; therefore, everything on the planet has its own heat limitation. Damage, destruction, and death are the natural outcomes whenever that heat limit is exceeded... simply, as humans, it doesn't take much heat to first hurt us, and then quickly kill us.

Thermal imaging cameras (TICs) are a modern tool that is currently being used to make smarter and a lot safer operational decisions about where the "red" is actually located and where it is going. Fire departments must develop local SOPs for how they will be integrated (dynamically) into the initial and ongoing situation evaluation process.

Command Safety

Situation Evaluation

% OF INVOLVEMENT:

The IC must determine the relationship of the involved and uninvolved fire areas. Making a determination of the percent of involvement is a practical way to establish this relationship. The IC must then combine the percent of involvement with the fire stage, along with total incident time, to develop an ongoing "moving picture" of current and forecasted incident dynamics. This "movie" will become the decision-making and operational basis to determine:

- ☐ the location, rate, and amount of changing conditions
- ☐ overall strategy IAP (tenability)
- ☐ incremental/ultimate outcomes
- ☐ incident time-line profile.

The IC must continually monitor and manage these factors to determine changes that quickly go from tenable to untenable. The IC

% of involvement Condition

1. 10%
2. 20%
3. 30%
4. 40%
5. 50%

% of Involvement (contd)

must always anticipate and build operations around the possibility of interior conditions, quickly becoming untenable. This reaction should become a regular part of how our attack is advanced and backed up. This reaction should include the quick application of accountability and rapid intervention procedures, calling for and maintaining a tactical reserve, quick sectorization, and the early assignment and support of safety officers. Responding to situations where conditions quickly change, in a way that threatens firefighters, is a lot easier for an IC who has developed the command and control habits that streamline going from plan A to plan B, because they automatically set it up and practiced it on the last 150 incidents. This IC evaluation, decision, and reaction process becomes very exciting because all the fire stages are not created equally, and they very probably are not evenly spaced along the standard time scale... simply, changes in the stages do not generally occur in equal time frames. Some changes are faster, some are slower... so some stages last longer, and some occur more quickly. Lots of times, the early stages that occur when the fire is smaller can last a long time, and then when the fire "picks up steam" (literally) and becomes more deeply seated, it changes (gets bigger), and many times it changes very quickly. This can present a huge safety problem when the earlier stage, which is at a smaller level and goes on for longer period at that level, allows interior

% of Involvement (contd)

attackers to enter, move around, and get deep into the building, and then fire conditions escalate quickly, and wreck (sometimes sadly end) the firefighter's day. While commercial buildings can get us in deeper, simply because they are bigger, we should not underestimate smaller (like residential) fire areas. Most firefighter structural fire fighting injuries and deaths occur at house fires, based on the frequency of these events, and the fact that the little ones produce exactly the same deadly conditions as the big ones. Wrecking your day being physically trapped in a burning clothes closet is not that much different from being in the same trapped position in a burning warehouse. The IC must develop the instinctive ability (and habit) to maintain command and control of structural fire operations that are being conducted on early-stage incidents. The IC must always predict the safety/survival challenges that the later stages of the current fire will present to the hazard-zone troops, and then begin to automatically set up the protective systems ahead of their need. This requires the IC to develop the ability and the approach to always be managing "ten minutes from now." These "nothing-showing/light smoke showing" situations simply have the most space left to expand, and such expansion can surprise, trap, and beat up/kill attackers who are not adequately protected, and who are disconnected from adequate-sized fire streams, and tactically-protected exit ways. The IC must always be in a position (starting from

% of Involvement (contd)

the very beginning of the incident) to effectively predict and manage changing conditions. This requires the IC to:

- ☐ never trust any incident condition to remain stable
- ☐ always be "awake" and fully conscious
- ☐ continually "read" the situation
- ☐ quickly exchange information/ stay connected by exchanging condition/progress/completion/ exception reports (read the gauges)
- ☐ be quick and agile.

Being able to determine and then track the percent of involvement becomes a major part of how the IC produces and pays attention to the "incident movie."

Note:

Using gauges to describe incident conditions is a new idea (at least to the authors and our command team). This approach is designed to be used by the IC in a fast-and-dirty kind of way, under pretty difficult conditions, so if it works, it can't be "rocket surgery" (as the "B" shifter said). We took the best shot we could at attaching numbers to the one-to-five part of the gauge to describe the various levels of intensity for that condition. There (obviously) isn't anything about the numbers that is

% of Involvement (contd)

magical... the percent of involvement is an example of that approach. We "pegged" the gauge at 50%, based on the reality that when half the fire area is involved, we are approaching (or well past) defensive conditions. The reader should apply their own local conditions, experiences, capabilities, and limitations to "localizing" their gauges by plugging in numbers that match that local tactical picture. Hopefully, the gauges provide a very basic and understandable way to package up and present critical fireground factors in a simple form where we can discuss, teach, argue, reflect, and better learn how those critical conditions affect the safety and survival of our troops and our customers... perhaps if the gauges help us to understand the details and dynamics of critical incident factors, they may also cause us to remember some of those details during actual incident operations.

Command Safety

Situation Evaluation

Smoke Condition

1. faint
2. light
3. moderate
4. heavy
5. zero visibility

SMOKE:

The original caveman figured out pretty quickly that the old homestead stayed a lot nicer if he built the fire outside the cave. He quickly noticed that the yucky part of the fire (products of combustion) magically went up and disappeared into the sky... 25 million years later, it took a complicated, very expensive, double, blind, longitudinal, seasonally adjusted government research study for some highly-trained scientist to discover that heat rises. Fires were originally designed to keep us warm, provide cheerful camp fire light, and to cook our roasted yakburger. Now we are learning that big-time problems occur when we build a building around the fire, as the fire progresses and fills the "cave" up with those same yucky products of combustion. A big part of the problem is that the same building features (walls, roof, doors, windows, weather proofing, etc.) that are designed to keep the elements outside now keep the smoke, heat, flame, and fire gases

Smoke (contd)

inside. When humans (customers/ firefighters) get trapped in the interior, and are now up close and personal with the fire products, it really gets ugly because breathing smoke really ruins the day of any air breather (like humans). When all this prehistoric stuff happens to a building and its occupants, we must now somehow deal with this "fire inside the cave" timeless process... taking on a timeless problem is always serious.

Firefighters must continually deal with smoke (and its effects) as they operate on the fireground. Nothing about smoke is nice. All smoke is toxic because it contains the poisonous and toxic products of combustion. If we breath it, it will first injure us, and then if we continue to breathe enough of it, it will kill us. To further complicate our lives, it is also flammable--add the right amount of heat (which sometimes occurs in seconds) and it becomes burning/ exploding smoke. It also obscures our vision as the fire burns, so it becomes a huge visual barrier to us being able to see, so that we can safely and effectively move around the fireground (particularly inside the fire building). Whenever we can't see, we can't evaluate and effectively react to what's going on. The level of visibility (or lack of) is always a major safety factor. Operating in dense smoke causes us to become disoriented and get lost. When this happens, we are in effect trapped by the smoke. Interior bosses (company and sector) must always be aware of smoke

Smoke (contd)

conditions and pay attention (continually monitor the gauge) to determine whether those conditions are getting better or worse. In areas much bigger than smaller, and the very beginning of medium, officers must regard serious smoke conditions as being very dangerous, particularly where those smoke levels are worsening. In these cases, when smoke levels reach four to five on the gauge, it becomes very difficult to conduct effective interior operations, particularly in large areas. Everyone on the team must operate with the very primitive understanding that when we are operating in smoke, our life expectancy(!) is absolutely determined by the time left on the air supply in our SCBA bottle. In these cases, when our troops lose their capability (for whatever reason) to exit the hazard area within the (very personal) time left on their air supply, they basically become smoke inhalation victims, casualties, and then statistics. We don't need to become smoke scientists to understand the smoke management problem--our ability to extinguish the fire on the inside, or to perform ventilation (in all forms), are very offensive-oriented operations. If we miss that typically narrow window where offensive conditions are present, and when the fire is continuing to burn and fill up the interior with smoke, we must regard this situation as quickly becoming defensive. We must consider smoke as serious a safety hazard as heat and structural collapse, because they are all nasty partners--first the smoke gets us, and then the other two take over and finish the job.

Command Safety

Situation Evaluation

STRUCTURAL STABILITY:

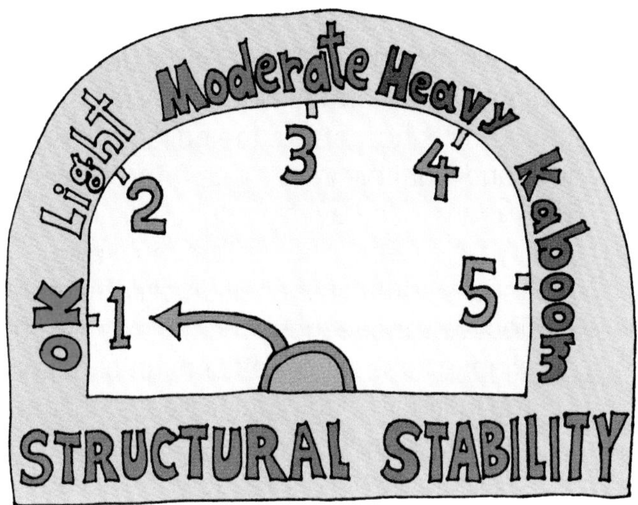

A building stands up because all the connectors that connect it together stay connected.* When that structure is on fire, the fire burns around and through the connections, and when those places that hold the building together become disconnected enough, gravity takes over, and it (simply) falls down. This is a big problem for anyone and anything inside and around the building--particularly for humans. When the building falls on them, it mutilates their body parts which always wrecks their day. This is an ever-present and timeless major occupational hazard for firefighters, who are generally the only ones inside a burning building (doing their jobs), because all the normal people ran outside twenty minutes ago, and have either gone home, or to the closest bar to celebrate the

* The reader is challenged to make up a sentence with "connect" in it three times!

Structural Stability Condition

1. OK
2. light
3. shaky
4. weak
5. kaboom

Structural Stability (contd)

fact that they became a drill press operator, and not a firefighter. The IC must develop and apply an information and reaction system that predicts how stable the building is right now, and what the immediate future looks like with regard to the ever-present building vs. gravity tug of war. The system must also continually control the position of firefighters to effectively move them out and away from collapse zones. Fire officers must develop the instinctive ability to look at stuff that is standing up and visualize where it will all land, when it falls down. This approach is based on basic gravity, and the geometric trajectory of such falling stuff. Where the stuff will land becomes the template for the collapse zone. This zone must be managed in a standard way by the IMS. Increasing numbers on the structural stability gauge must be connected to increasing security in the collapse zone and

Structural Stability (contd)

into the collapse zone. The IC must react to structural stability signs like sagging, creaking, leaning, goofy noises, bulging, smoke, and water coming out of places where it's not supposed to come out of, any place where we have applied a lot of water (very heavy), or any time/place where the fire has burned ten to twelve minutes, particularly in concealed spaces like attic areas that expose roof systems. Throughout this essay, we talk about the seven sides of a structural fire situation: top/bottom/inside/four sides AND the building layers that go with each side that are essentially the concealed space that goes with or is attached to that "side." Huge safety surprises (sometimes fatal) occur when fire hides in these spaces, and either breaks out unexpectedly, or causes the structure to collapse. Everyone on the command and operational team, all over the incident, must identify such concealed spaces, open them up if they suspect they are involved, operate fire streams into them if they are involved, or stay away from them if they can't gain access and take hydraulic control of them. The IC must continually watch the structural stability meter and begin to think defensive at about 2.5 and be moving the troops outside and away at 3.0 or so. Our PPE will protect us from some lightweight, interior-finish stuff falling on us (plaster, drywall, etc.), but no part of our standard safety system can protect our troops from

Structural Stability (contd)

actual structural collapse. Buildings are very heavy; there is really no such thing as a minor collapse. Our bodies do not hold up very well (or really, at all) to structural stuff collapsing on us, or under us.

Command Safety

Situation Evaluation

FIRE LOAD:

Simply, fire load is a big deal because it is what burns. Structural fires burn through the "numbers" from just having smoke present (1), to a light contents involvement (2), to a heavy contents involvement (3), to burning the interior finish (4), on into structural involvement (5). Big load = big fire/not so big fire load = not so big a fire. Not rocket science, but we better evaluate how the fire has affected the fire load and then operate accordingly. We must pick the correct "artillery" to attempt to control what is burning. We must be careful of developing a pea-shooter mentality, because agile little shooters (such as 1 1/2"/1 3/4" attack lines) work so well in lots of our repeat (residential) jobs. We get in trouble when we habitually take these little pea shooters into big battles. The determination of the amount, nature, and arrangement of the fire load is a major factor the IC must establish, and then develop a water show that overwhelms the BTU output (rate of flow). If we can't develop such a big enough/fast

Fire Load Condition

1. light
2. light +
3. moderate
4. moderate+
5. heavy

Fire Load (contd)

enough/in the right place interior attack before the fire "takes over," then we must go defensive--which means we basically get in between the fire and the exposures, and then let the fire load burn itself out, while we set up and operate an exterior defensive hydraulic attack. The IC and the command team must realistically evaluate and forecast fire progression (time/temperature), and continually control the position and function of the troops, in relation to the profile (amount, nature, and arrangement) of the fire load. Sometimes, this means giving the fire the biggest, quickest interior shot we can produce, evaluating the effect of that attack, and if conditions get better, keep going offensively--if the fire gets bigger, abandon the interior and set up outside defensive operations. Burning fire load is a function of the basic combustion process which is just an out-of-control physics problem (in relation to our safety/fire control capability)--nobody can repeal the scientific laws of physics (that's why they call 'em laws and not suggestions)... but when it is necessary, we can stay out of the fire's way, not get it on us, and marvel at the magic of combustion, from the outside (based on our understanding of and experience with these laws). It is critical that we have a standard fire-load scale so that we can effectively exchange current, critical information regarding the size of the fire problem, and our ongoing ability to find the fire, cut it off, and put it out. Building fires occur along a fairly standard scale, with a progressive series of stages. If we connect the "snapshots" of each separate stage together (from beginning to end), we get a

Fire Load (contd)

complete picture of a fire event and a description of what happens to burning buildings. Building fires typically start small, grow to a rate that involves all the major portions of that fire area, and then burn down to a point where the fuel is all burned up. The IC must superimpose the find it/cut it off/put it out process along the standard fire progression scale (that we call the time/temperature curve). Manual fire fighting is (obviously) most effective in the beginning of this process--but sometimes the manual fire suppression stage misses (for a lot of reasons) the front-end fire control opportunity. The period when the practical and safe offensive chance to control the fire from the inside goes away can be pretty brief. This becomes a critical moment for command, control, and safety. An effective evaluation of the fire load, along with a realistic prediction of the immediate future of our fire control success becomes the basis of how long the IC conducts offensive operations. When the IC determines that the offensive strategy is "shaky," and the operation is now creating a big risk and is only protecting property (rescue completed), it is time to move the troops outside to defensive positions. When this occurs, command and control must now go from a written procedure (SOP) to an actual practice (show time).

Command Safety

Situation Evaluation

Occupancy Hazard Condition

1. OK
2. light
3. moderate
4. heavy
5. ultra

Occupancy Hazard:

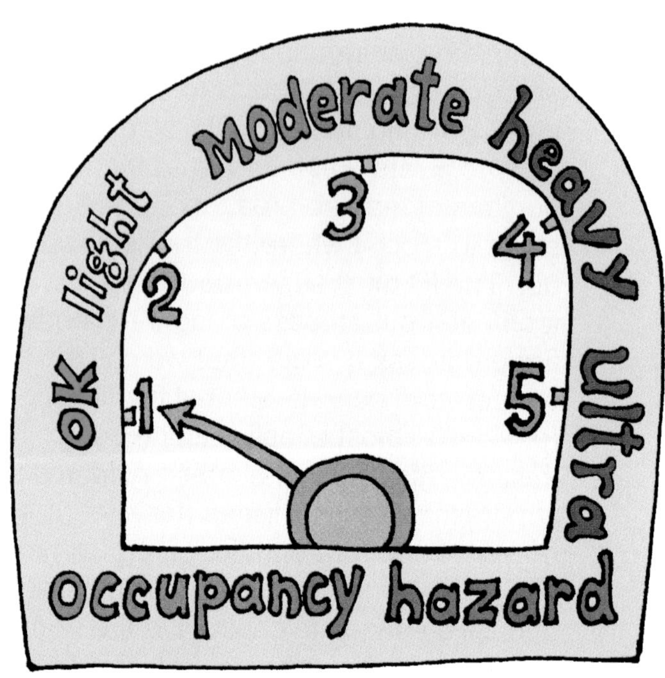

The type of place that is "hosting" the fire has a big influence on how that fire will occur. The type of occupancy inherently creates it's own set of characteristics that will effect how the fire starts, extends and sustains itself, and the special tactical problems that firefighters will encounter and must overcome. Occupancy hazard involves construction, age, arrangement, contents, process characteristics, physical barriers, and special problems. Lots of those characteristics are generic to that type of occupancy. Just reporting "church," "school," "plating shop," "grocery store," "bar," "barn," or "house" (as examples) provides a basic picture of what thc place looks like, how it is used, and the typical conditions to expect. When the place is in fact typical, just reporting the occupancy type is an effective form of communications shorthand--reporting exceptions to the typical

Occupancy Hazard (contd)

profile (i.e., age, size, condition, etc.) begins to expand the description. The IC must add up all the critical factors that relate to the occupancy to begin to develop a profile on the degree of hazard the place presents. The IAP must be developed in a way that addresses the details of that hazard. The more the IC and the team knows about the details of that occupancy type, and the history of the operational experiences that have occurred in those places in the past, will make the response safer and more effective. The operational plan must consider the level of occupancy hazard and then create a matching response to meet (or exceed) that degree of hazard. Lots of firefighters have gotten scuffed up (badly) because they thought they were fighting a lightweight opponent, and they suddenly discovered they were getting thumped by a heavyweight.

Command Safety

Situation Evaluation

Residential/Commercial Condition

1. small-med residential
2. med-large residence
3. small-med commercial
4. med-large commercial
5. huge-ultra commercial

Residential/Commercial:

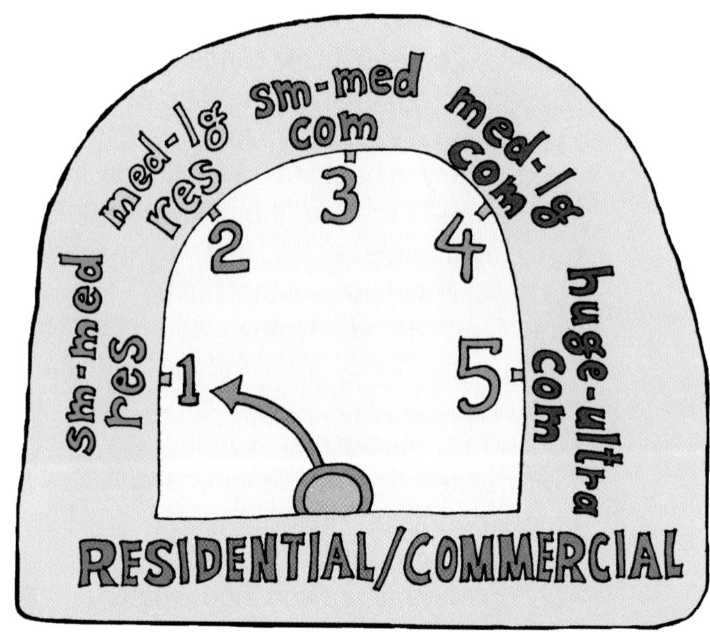

A major IC function is to create an operational and command system response that matches the size and nature of the incident. A major occupancy hazard factor involves how we approach, evaluate, and operate at incidents in residential and commercial occupancies. We mostly attend and operate at single family residence fires, so we (naturally) develop single-family command and operational habits. The safety challenge presented by these habits is that there can be a huge difference between residential and commercial fires. While residential fire situations are sufficiently dangerous to challenge our basic safety system, they are much more forgiving than larger, more complex commercial incidents. We can recover more easily on residential fires then we do on marginally unsafe basic attack operations, where we routinely exit the hazard area with

Residential/Commercial (contd)

zero SCBA air levels, operate at the very end of the safe attack distances, and perform sloppy attack line/exit marker operations. After going to a bunch of house fires, our ICs can stay half asleep with their head partially disconnected, and the troops will almost automatically solve the incident problems (yawn). Over time, we habitually practice and refine breaking these basic safety rules so much they start to "look okay" to us. Then on a dark and windy night, we encounter a working offensive fire inside a large commercial building with a big-time fire load. Now it's judgement day for all our bad habits because now we are operating in a place that will quickly out perform the day-to-day systems (and habits) we have become comfortable with. Simply, the commercial place is too big for our day-to-day single-family residence approach. We have to go too far to get to the problem (which is lots bigger than the routine house fire room and contents). Most of the time our commercial command operational response is too little/too late. As we have discussed in earlier gauge descriptions, most departments have a refined, practiced, safe DEFENSIVE commercial standard fire fighting plan and approach... basically, what we do is surround the building with fire streams and patiently apply exterior master streams as we (safely) protect exposures and witness the involved fire area burn down. In many defensive situations, the extent of the fire makes the strategic decision for us and we just cooperate with those

Residential/Commercial (contd)

expanded conditions and operate from the outside--the absolute correct action. This is an example of connecting a standard response to a standard condition, to achieve a standard outcome. The problem is that most of us do not have such a well-practiced commercial structural fire fighting plan for an OFFENSIVE situation. In commercial building fires, we must describe realistically both our actual capabilities and limitations, and then respond and react within those local realities. We must also practice quickly calling for additional resources along with escalating our regular systems to "fit" commercial situations. That escalation must include a large and early execution of the following standard system elements:

- ☑ command team and staff
- ☑ sectors
- ☑ RIC teams
- ☑ accountability
- ☑ safety officers
- ☑ rehab
- ☑ attack line management
- ☑ initial and ongoing exit way/exit control
- ☑ support activities
- ☑ overall deployment management
- ☑ resource assignment/inventory/tracking

An IC who is engaging in offensive interior fire fighting operations should conceptually think of and approach sending those troops inside a structure, as if it is a (really large)

Residential/Commercial (contd)

confined-space operation. When we go from the outside to the inside of a structure, the building (remember the "cave") creates an enclosure that physically holds the products of combustion inside. Those products of combustion (heat, flame, smoke, fire gases) almost instantly begin to damage, destroy, injure, and eventually kill everything and everybody who is inside with the fire. This is why ventilation is such an important support activity to offensive operations (God bless the ladder/truck companies). In typical confined space rescue situations, the customer is being held captive by the confined space in a way that requires physical rescue by the responders. Many times, the confinement conditions are very hazardous and require the IC to establish extensive safety and support services for the rescuers... stuff like big-time RIC teams, strong communications, lots of safety officers and bosses, special operations specialists and technicians, backup equipment, reinforced air supply systems, operational sectors that typically over manage the hazard zone, etc. We take whatever time is necessary to set up and then micro-manage the standard operational and safety system before (and after) we go into the confined space.

When we conduct regular offensive fire fighting operations, we must extend our operational support and safety systems into the hazard area just like the confined space incident. What is different in most fire fighting operations is that the area we are going into

Residential/Commercial (contd)

is typically not confined to begin with (in fact, it is generally just the opposite--it is expansive). The size of the interior (i.e., the size of the building) creates its own set of safety challenges. Most of the time in a confined space situation, we know where the trapped customer is, and we don't get lost going in to deal with that customer. We are also able to provide a well-supported prepared path (entry and exit) that leads to the customer, and we provide the required support so that the path is highly protected. What happens at a structural fire is that we almost instantly enter the hazard zone, as quickly as we are able. We then "take on" the entire inside of the fire building to complete the standard tactical priorities. Our regular safety system provides the protective capability that allows us to move around inside the involved fire area which many times is full of the products of combustion. Based on our mobility capability, the larger the building, the farther we can go (duh!). This offensive routine is okay until something goes wrong, and what goes wrong is not a mystery (because it has gone wrong for 250 years): collapse/ thermal-toxic insult/lost/trapped. When one (or more) of these hazards grabs our firefighter, now we have almost exactly the same rescue challenge that occurs at a confined space incident--except, many times we don't know where that firefighter(s) is in the (many times large) building. They are generally low on air or out of air, and they are surrounded by the products of combustion. At this desperate point, it becomes

Residential/Commercial (contd)

very difficult for us to set up the rescue systems (almost identical to what we do at a confined space operation) in time to locate, stabilize, and remove our firefighter.

The North American fire service has made a significant investment in discussing, developing, training, and doing rapid intervention. When we actually examine what goes into rapid intervention operations, it is a task-level activity. When firefighters who are operating inside the hazard zone become lost, trapped, or missing, our rapid intervention plan and reaction is designed to send more firefighters inside to get them out to safety. When firefighters become "entangled" in the hazard zone, the choice is pretty simple. If it is humanly possible to get them out (rescue them), that is what we are going to do. This is the most desperate, high-risk situation any IC will ever face. When firefighters become lost, trapped, or missing, it is because some piece or part of the regular safety system has failed (our incident management, safety systems, and SOPs are designed to keep firefighters both effective and safe--we all go home okay when the event is over). Often times, Mayday situations are the result of changing conditions. In these instances, the RICs that the IC deploy will have to operate in the same

Residential/Commercial (contd)

conditions that caused the initial crews to get into trouble. This is truly last-chance, crisis management. The primary intent of this book is for the IC to do the regular, reoccurring, strategic-level management (i.e., the functions of command) that are automatically performed at every incident to reduce and hopefully eliminate these types of task-level, possibly catastrophic events from ever occurring.

One of the operational realities of RIC operations is that, in many cases, it is not rapid. The farther firefighters are from the building's exits, the longer the RIC operation will require. In the most instances (based on a large number of drills in commercial buildings), it will take in the neighborhood of twenty minutes for the IC and RIC crews to develop a RIC plan, deploy into the structure, locate the firefighters who are in trouble, and get them to safe locations--this is in a medium-sized structure (less than 12,000 square feet). It will also require the efforts of multiple RICs to pull this off. The major biological and operational challenges that we must consider in these types of incident operations are the very limited amount of air we enter the hazard zone with (less than twenty-minutes working time in most cases) and how long the building well remain standing while being eaten by an active fire.

Establishing and maintaining the capability to do RIC operations will always make hazard-zone operations safer. This capability must be tempered with a strong dose of reality.

Residential/Commercial (contd)

Just the presence of a RIC team can never become a reason that the IC uses to keep firefighters in positions that are being adversely effected by worsening conditions. It is more effective, safer, smarter, saner, etc., for the IC to base the strategy on the actual (and forecasted) conditions, the tactical priorities, and the risk management plan. If the IC assigns RIC teams to certain locations, based on their concern over the changing conditions in that area, the question the IC must be able to answer is, "Is the risk to my troops worth the possible gain?" We will only take a big risk to save savable lives. If fully-protected firefighters are pushing the outside limits of our standard safety system, the unprotected (customers) have been lost. It is time for the IC to go defensive, to get PARs from all of the hazard-zone workers, establish collapse zones, flow big water, and order food for the troops. This is how the IC matches a standard approach to standard conditions, keeps the troops okay and alive to fight another day (standard outcome).

Our local IMS applied to fires in buildings must respond to this cheerful picture in two ways:

1. We must change how the IC manages structural fire fighting operations, particularly on the strategic level to PREVENT the need to rescue firefighters. This entire manual is directed to making that adjustment. Many confined-space rescues are really not rescues, but are body recoveries and, sadly the same

Residential/Commercial (contd)

applies to the rescue of firefighters at structural fire operations. The best way to get out of trouble is to not get in trouble.

2. We must add a set of practical, doable operational (tactical) procedures to our regular structural fire fighting routine. These changes will serve to both prevent safety trouble and will also respond to such trouble when it occurs. A lot of these activities begin to make a structural fire look at least a little like a special operations incident. The IC must lead this change by managing their role on the strategic level. The focus of this manual is directed toward making these changes.

Our service has done our interior fire fighting routine for the past 250 years. We are the best hazard-zone operators on earth. We have and we will continue to keep our hazard-zone promise to Mrs. Smith--simply, we will go in and do everything we can to get her out when she needs us. The problem is that we have done it so long and so much that we have lost a wee bit of respect for the hazard zone. The hazard zone loves it when we underestimate its "hazardness"--it just bides its time, picks its spot, and kicks our ass. We do not need to suffer a huge personality change to make these adjustments... a small change in our operations can produce a huge change in our safety.

Command Safety

Situation Evaluation

Access In:

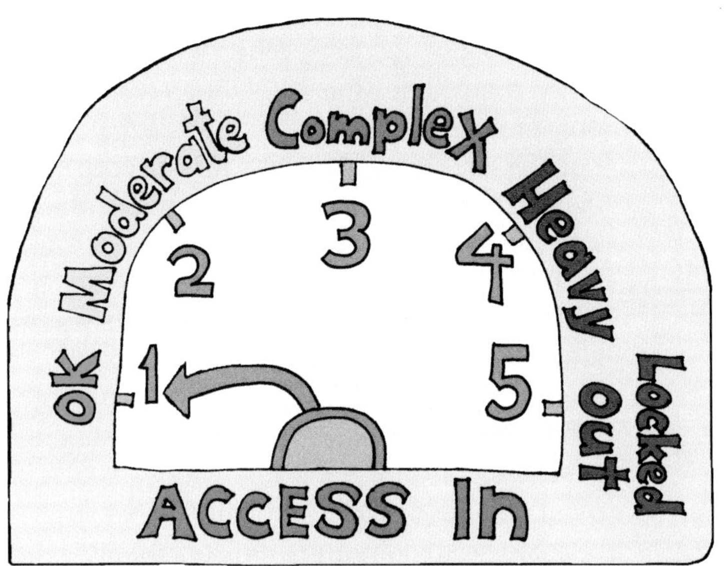

Our lives can become extremely complicated when we encounter fire/buildings/areas with a lot of built-in or natural security, obstructions, or other access "confusion." Finding our way into or forcing entry into these places can take a lot of time and effort. When this occurs, the fire is typically getting bigger and becoming more deeply seated (i.e., burning into the interior finish and then into the concealed spaces), so the IC must consider where the fire will be when actual entry is achieved. Another problem is that in many of these cases, there is the same high level of security on all the yards, access areas, and building openings (doors/windows)--if there is a ten-foot, chain-link fence with three rolls of razor wire on the top, being patrolled by a pack of howling, hungry, junk-yard dogs, and the front door is built like a safe, there is a good chance that the rest of the place has the

Access In Condition

1. OK
2. moderate barriers
3. complex entry
4. heavy security
5. locked out

Access In (contd)

same level of security. When we do gain entry, many times this is (at that point) the only entry and exit place we have into, and out of the building. Now the IC must be aware that we are in a one-way in and one-way out situation, and there is no "back door" that can provide an in/out alternate. In most cases, hose crews have been anxiously standing by, waiting for the trucks to bash their way in, and when an opening is made, they are in a huge hurry to get in and attack. In such single entry-point situations, the IC must understand how labor intensive this building bashing is and then continue to produce the resources (and orders) to quickly provide as many interior entry/exit points as possible. The need to perform such access/entry manual labor becomes an excellent reason for the IC to strike additional alarms. Such additional alarms should be struck sooner, rather than later. The IC must also quickly protect and "harden" the single-entry point with lights, guide/usher crews, hose lines and RICs. The IC must alert companies and sectors that there is currently only one access point available and to operate accordingly. Big-time, over-engineered fences, gates, grills, doors, and dogs don't know the difference between the bad guys and the good guys--it's a tough deal when the good guys finally get on the inside and quickly need to get out, and now the building security is doing just what it was designed to do. Other points of entry/exit should be created and also "hardened" as

Access In (contd)

quickly as possible. The IC must pay attention to completion reports when another entry/exit opening is made into the hazard zone. The IC must then communicate to sectors and companies that another access point is available and must integrate those entry/exit points into the IAP.

Command Safety

Situation Evaluation

Exit Out Condition

1. OK
2. complex
3. detained
4. stuck
5. fat-assed trapped

Exit Out:

The IC must continually consider hazard area "geometry" and then set up and manage interior operations around how that area is arranged. A major part of that consideration involves maintaining as many clear exits and exit ways out of the hazard area as possible. This is a lot easier said than done, simply because there are a lot of things that can mess up the ways we use to get out of a building that is on fire. The first is that the building quickly gets filled up with the products of combustion that obscure visibility, creates a superheated and toxic environment, then the fire involves the contents, then the interior finish, and then the structure itself and then the fire begins to demolish ("disconnect") the building and to cause it to structurally come apart and to begin to fall down (very messy). All these exciting things can make it pretty tough to find even the swellest, big, open exits. We

Exit Out (contd)

typically take in hose lines that we use to provide a water way in and then use them to mark the pathway out for exiting firefighters. Major problems occur when multiple hose lines perform their normal "loop de loop" and end up crossed over each other, and all tangled up ("spaghetti"), so they become difficult to follow out. It can be fatal to get off and lose the hose line in a high-hazard situation. When we get disoriented and lost in the products of combustion, our life expectancy gets quickly and directly connected to the air in our bottle. We must always understand and be aware of how inherently risky interior operations are under fire conditions, and the harsh reality that we are actually betting our lives on staying on a hose line that can (and does sometimes when we get off the line) quickly become a needle in a hay stack. The IC must have the personal fortitude to regard gauge readings that reach about the middle point (big 2/small 3) in some places (big, complex, bad access, congested, etc.) as simply being too dangerous to put crews into under fire conditions until we "prepare" that building for interior operations. Such preparation involves the early call for the resources required to create and maintain "tactical corridors." The IC must understand these realities and provide protection for these exit ways in situations where interior operations are being extended into buildings.

Command Safety

Situation Evaluation

Exit Out (contd)

The tactical preparation of fire fighting exit ways must involve activities like:

- ☐ quickly establishing multiple exit points
- ☐ advancing attack lines into critical (exit) areas for both fire fighting and to maintain control of interior access
- ☐ quickly calling adequate (operational and tactical reserve/rescue) resources to sustain large, long-term operations
- ☐ assigning sector and safety officers
- ☐ providing guide crews that serve as "ushers" along/on hose lines to crews coming in and going out
- ☐ lighting up exits and exit ways early in their operations
- ☐ critical ongoing evaluation of the current strategy in relation to current and forecasted conditions.

An example of this tactical preparation of the fire building is when the IC makes "opening up" the entrance/exit points a top priority. It is a lot safer and ultimately quicker to utilize/ assign resources to open up the multiple doors on the four sides of a 7,000 square-foot building than it is to assign those same

Exit Out (contd)

resources to pile through one door and stumble around in level-three smoke. A big-time, happy outcome of the IC creating multiple entry/exit points is when one of the forcible entry crews radios the IC that they have located the fire fifteen feet from the door (they just opened) and that the fire can be controlled with a single line... rather than dragging attack lines 250 feet through the products of combustion from the single-entry point on the opposite side.

Command Safety

Situation Evaluation

Interior Arrangement:

Interior Arrangement Conditions

1. OK
2. confused
3. congested
4. obstacle course
5. grid lock

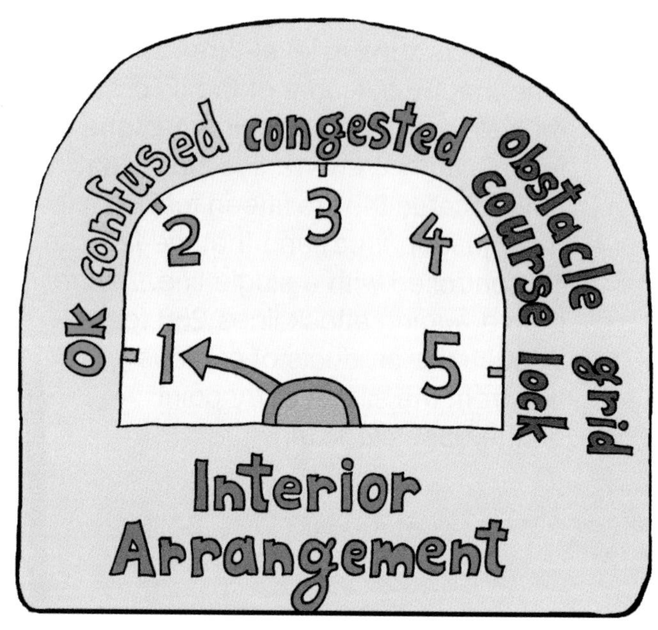

The way the interior of a fire area is "loaded" (amount/location) becomes a big deal in how we can get into that area, effectively operate and then (most of all) safely get out of that same area. Interior operations are closely connected to the arrangement of clear spaces, aisles, clutter, shelving, contents, storage stability, junk, stock, housekeeping, etc. As the interior area becomes more congested, we quickly lose our mobility and access options, and it becomes more dangerous to operate inside, and then to exit safely back outside. Interior crews must quickly evaluate the degree of interior complication, and as the fire loading and the limitation of interior movement goes up, they must adjust how (and if in some cases) they go into and move around, and maintain an

Interior Arrangement (contd)

ongoing capability to exit that area. In some cases, exit routes must be protected, reinforced and "hardened" with hose lines, lights, and crews assigned to maintaining multiple in and out access points. Interior crews must always predict what will happen if conditions worsen, particularly in complicated areas where fire conditions are light and allow easy entry and movement. Situations that were easy and quick to get into, and then quickly become impossible to get out of, have killed lots of firefighters. We should think of interior congestion and obstructions as a one-way check valve. The valve will let us in, but after we get in and conditions quickly change, it won't let us come back out. Remember, deciding where you can go (on the fireground) must always include a conscious evaluation and decision about if you can get out of that position, particularly if the interior conditions in that place quickly worsen--just because you can go somewhere, doesn't mean you should. Every smart interior firefighter with any seniority got old because they developed the natural ability (and habitual inclination) to look forward once and look backward twice. Another major IC situation evaluation challenge is to use all the information sources to determine the basic physical layout of the involved and exposed incident areas. The basis for this determination involves the fairly standard structural fire "geometry" that involves the seven sides of a typical building, the concealed space area "layers" that are attached to each side. The seven sides = inside/top/bottom/side one/side two/side three/side

Interior Arrangement (contd)

four. The entire team must quickly get recon information on all seven sides, including a report on access and mobility in that area to determine initial involvement and to predict where the problem (i.e., generally fire) is going. Everyone must be very careful of any side/layer condition which is unknown. These unknowns must become critical information targets. The IC must account for, protect, and track crews while they do recon. Crews must physically evaluate (in some cases, penetrate) and verify the condition (if fire has extended into that side) of the concealed-space layers on all seven sides:

- ☐ attics
- ☐ basements
- ☐ multiple ceilings
- ☐ concealed spaces
- ☐ construction voids
- ☐ utility shafts
- ☐ duct work.

The IC must set up to continually evaluate ongoing changes in conditions on each side to avoid surprises, and must establish and maintain control of the assignment (position/function) of the troops and to call for and keep available a tactical reserve to maintain the capability to respond to changes. The IC must realize that a radio report from one side is a description of only one seventh of the entire structure. The IC must also determine the basic shape and elevation of the building to understand how the seven sides are arranged. Sometimes buildings are a big

Interior Arrangement (contd)

square box with just four sides, other times they look like a jigsaw puzzle. Also, the elevation can be real simple when the building is built on flat ground and looks (and is) the same from every side. In other cases, the place is built on a hill, and floor one in the front is floor three in the rear. The complexity of the geometry will challenge the IC getting an accurate picture of the fire area--the more accurate that picture is, the safer the operation will be and the less chance an "architectural" surprise will hurt a hazard zoner.

Aggression Condition

1. coma
2. moving
3. controlled hustle
4. moving quickly
5. running berserk (pell mell)

Aggression:

Most people believe and admire that firefighters are naturally fast and active as they approach and operate on an incident. When we become a firefighter, we get aggression drummed into us as a big-time, peer-driven virtue, starting our first day as a recruit. A reflective, cautious new comer would not get very high ratings. In fact, their recruit training officers (RTOs) would probably suggest such a person seek some intellectual pursuit, like being a philosophy professor, tide watcher, or toll taker. How fast we move and where that movement takes us on the fireground becomes a major safety and survival factor. Like every other virtue, aggression must be mixed with a variety of other skills, traits, and abilities so it can be applied in a versatile manner along a scale,

Aggression (contd)

based on the conditions that are present. We must be careful that acting aggressively (based on our early training and continual reinforcement) does not become a post-hypnotic suggestion that becomes automatic. Most of the time, the fastest way to operate on the fireground is to slow down (a bit), look at what is going on around (and ahead of) you, plan where you want to go, and how fast you must get there--most of all figure out the reason you are making the trip into the hazard zone and what you must take with you. We must develop and apply a natural routine, so we can safely get the job done, and then come back out (okay). This requires we learn, apply, and refine a thoughtful and conscious approach to how we do our work--particularly what we do in the hazard zone. This does not mean that we appear timid or tentative. During normal incident operations, we must send the message to our customers that we are being responsive to what is (to them) an urgent event. The middle place on our gauge (level "3") is connected to a level we describe as a "controlled hustle." This combines the appearance of us moving actively to solve the incident problem, along with us being effectively under control. This operational speed is not too fast and not too slow. Obviously, if there are walls falling, we quickly move the speed to a "5"... if there are folks shooting at each other as we arrive, we set the gauge on "0" (or even minus, if we must retreat). A big part of smart, young firefighters gettin' to be smart old firefighters involves becoming skillful in

Aggression (contd)

determining what level of aggression is effective and safe in each separate situation. Doing this becomes the practical basis for how we pace ourselves.

Command Safety

Situation Evaluation

IC's Instinct:

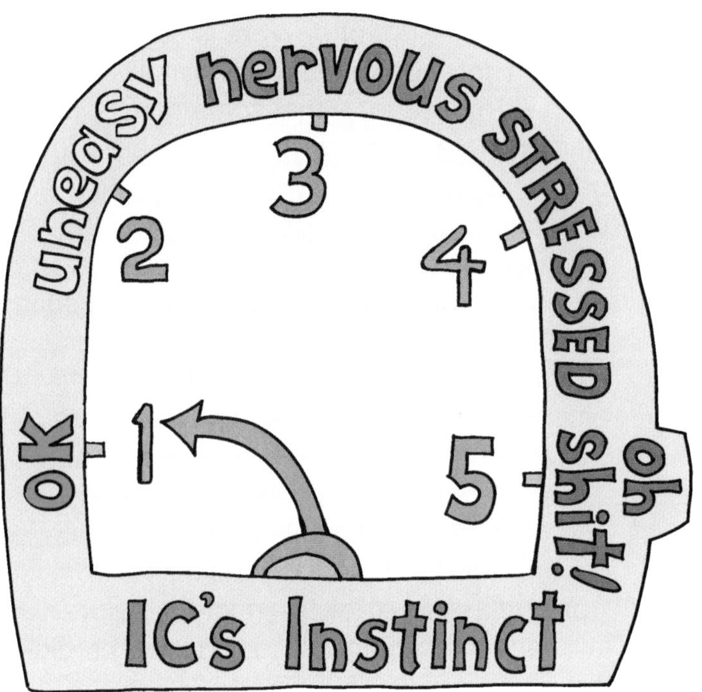

Sometimes there is some non-definitive thing about an incident that the IC is just not comfortable with. The IC doesn't know exactly what it is that is causing that "funny" gut feeling, or why the hair on the back of their neck is standing up. Many times, when this occurs, the IC is receiving a message that is connecting just below the conscious level. That "feeling" is sending the IC a subtle message that must be acknowledged, simply because most of the time the message has something to do with safety. This is when the IC must trust their instincts and react to the subliminal alert. When this feeling is followed by a sudden, unpredicted, unexpected and up until then unseen violent event, then the feeling is verified--the AC unit comes crashing through the overhead, the roof comes in,

IC's Instinct Condition

1. OK
2. uneasy
3. nervous
4. stressed
5. oh shit

Command Safety

Situation Evaluation

IC's Instinct (contd)

the fire blows up, etc. What is generally happening when we get this instinctive feeling (i.e., message) is that something is alerting our eyes, ears, nose, skin, brain just below our conscious level. Our training, experience, road rash, reflection (while we lay there in the dark watching the ceiling fan go around) all gets loaded into us on a bunch of different levels. Smart, crafty ICs pay attention to every level and quickly react to incoming messages that are many times packaged up in safety "sleepers" and "red flags" before they occur. Even a tactical idiot can tell five minutes after the wall fell, that it fell... the trick is to predict that it will fall five minutes before it falls, and then effectively react to that prediction.

Command Safety

Situation Evaluation

IC Checklist:

- [ ] Quickly identify and react to safety "red flags."

Safety Effect:

A major part of the initial and ongoing size up (reading the gauges) involves paying special attention to the conditions that create serious safety hazards to our workers. These "red flag" conditions are the hazards that are above and beyond the regular incident conditions that are generally present at structural fires, and are the conditions that can evolve beyond the hazards of our regular safety system protection. They can create a serious level of risk that requires us to identify and react to that condition as quickly as possible. The "red flags" emerge from the regular stuff that occurs during incident operations, but they generally involve some combination of regular factors, or some "quirk" of a regular condition. A big-time evaluation function of the entire response team is to pay attention to how the incident is occurring and to predict/eliminate/move away from safety problems. In their most severe form, the "red flags" are fatal to our firefighters. In their least severe from, an early (but escalating) "red flag" can be disguised as a "sleeper"-- these are sneaky demons that don't appear to be serious conditions at that moment (or stage), but if they are allowed to develop can become a major ass kicker. Being able to forecast the future of what current conditions can become is a major reason we blab and blab about how

Command Safety

Situation Evaluation

Safety Effect:

important forecasting is to safety. The ability to do such prediction emerges mostly from seeing the beginning/middle/end of a lot of tactical stuff that looked okay one minute, and turned into absolute "doodoo" the next (... this is why smart travelers always want a new plane and an old pilot... for us it's an old IC (old in wisdom, not necessarily age) and a new Suburban... very snazzy. The IC must know what these risk-producing conditions look like (particularly in their early stages), how they behave as they continue to evolve, and the correct action to take in dealing with them. Some of these "red flag" conditions can be managed (eliminated) by applying standard action, and some simply must be avoided. The IC must quickly react (act/avoid) to these warning signs to stop the "snowball" effect, i.e., the unmanaged hazard picks up speed, creates/attracts other types of safety problems, and then gets a larger and larger life of their own. At some point, it becomes unstoppable by any sort of normal operational action. Firefighters who get caught in the "snowball" generally get beaten up (or worse). Lots of times when this process occurs, there were early warning signs present, but the IC didn't see the signs, saw them and didn't "get it," or saw the "red flags" and didn't effectively react to them. Everyone operating on every level must be "red-flag" literate so they can quickly communicate the sighting of such a "red-flag" condition from their position (big deal).

Situation Evaluation

Safety Effect:

In addition to being "red-flag" literate, the hazard-zone workers must be empowered to effectively report and then to quickly react to "red-flag" conditions--this can range from making a small adjustment to a complete change in their position/function.

Red Flags:

This evaluation, identification, and reporting process is an absolutely critical operational function that consistently has life or death potential. Lots of our departed troops have unexpectedly gone on to the happy-hunting ground because someone saw a "red flag" waving and they just waved back, and didn't tell someone (mostly the IC) who could have moved the troops to safety. "Red flags" include stuff like the following:

- ☐ anything that is "iffy"... if you have to ask or wonder, assume the worst
- ☐ any free-lancing, i.e., action that was not assigned and is not being tracked, accounted for, monitored, managed, and backed up
- ☐ active fires on multiple sides of firefighters--particularly below/ behind them
- ☐ fires that don't react to standard (correct) attack action
- ☐ firefighters operating around active fire conditions that are not protected by a fire stream

Command Safety

Situation Evaluation

Safety Effect:

Red Flags (contd):

- ☐ large, complicated, congested floor areas
- ☐ one way in/one way out access situations
- ☐ any in/out access situation with lots of stuff stacked around it
- ☐ fires that seem too simple ("sucker punches")
- ☐ developing and then consistently operating with a "residential" mentality and approach
- ☐ fires that look like they are about to rollover, backdraft, flashover (lots of times they are)
- ☐ goofy, unusual, scary noises
- ☐ anything we do a lot that is extremely familiar ("repeat business")... can have a "sucker punch" hidden somewhere
- ☐ unnatural smoke behavior--color, density, travel, temperature, location
- ☐ lots of work to do with limited resources
- ☐ fires in concealed spaces, voids, walls, ceilings, construction voids, attics, truss lofts

Command Safety

Situation Evaluation

Safety Effect:

Red Flags (contd):

- ☐ large, open spaces, big, unsupported roof/floor areas with no upright columns
- ☐ fire is present but can't be found
- ☐ rapid build up of heat, smoke, fire gases
- ☐ change in "draft" directions--the wind
- ☐ fires that "won't go out"--particularly when normally effective resources and action are in place and operating
- ☐ extended time periods (ten min. +) of offensive fire fighting--ongoing offensive operations that are not improving
- ☐ anything that is sagging, leaning, twisting, or bulging that isn't supposed to be (very little is supposed to be)
- ☐ the ongoing effect of incorrect action, bad positioning of resources
- ☐ uncommanded fires (no IC/screwed up IC)
- ☐ unusually large amounts of combustible stuff

Command Safety

Situation Evaluation

Safety Effect:

Red Flags (contd):

- ☐ any man-made structural thing a firefighter must operate under that God didn't put there (air/clouds)
- ☐ extended time periods with no more/ better hazard information (info is getting worse rather than getting better)
- ☐ water/smoke/"stuff" coming out of/ through unnatural places
- ☐ poor/no ventilation
- ☐ improper tactical action
- ☐ conflicting reports (particularly if they are about the same place/thing)
- ☐ reports to the IC that don't match what the IC is seeing
- ☐ past events or conditions that look a lot like the current incident but are sufficiently different enough to produce a dangerous surprise
- ☐ crews are chasing smoke and asking for more help
- ☐ your gut tells you not to believe what you're hearing and seeing

Command Safety

Situation Evaluation

Safety Effect:

Red Flags (contd):

- ☐ when you prematurely assume the task is done (things got okay too quickly)

- ☐ the inside/outside commo link is broken. The IC calls and no one answers (radio)... or the inside calls and the IC doesn't answer

- ☐ assuming you have normal construction when it is (in fact) lightweight/substandard construction

- ☐ initial-attack lines have little or no effect on the fire

- ☐ any condition or event that causes the IC to lose track of the where/what/welfare of the crews

- ☐ nothing showing (or very minor) situations that "suck us in" and then change quickly

- ☐ any basically unsafe practice that becomes a habit and then starts to look okay

- ☐ any operational practice, procedure, or activity that is not routinely critically reviewed.

Command Safety

Situation Evaluation

Safety Effect:

Note

This list is only meant to offer examples of the safety conditions that can sneak up and zap us if we don't identify and react to them effectively. The reader is encouraged to add any others that occur to them and those they experience. It's always smart to keep a list of the dirty tricks a situation can pull on you--particularly when you're going up against a dirty devil who doesn't play fair.

Command Safety

Situation Evaluation

IC Checklist:

- ☐ Structure and time info management around tactical priorities and firefighter safety.

Safety Effect:

Wherever we can create well-known, highly understood operations, we get a big jump on safety, simply because everyone knows the game plan and can play their position in achieving the standard tactical objectives. A major way we do this is by structuring situation, evaluation, hazard analysis, and info management around the standard structural fire fighting operational tactical priorities (rescue/fire control/property conservation) and then creating operations to match that order to keep everyone working toward the same incident goals. This tactical priority-based coordination insures the entire fire area gets searched, the fire gets knocked down, all sides of the structure get checked for extension, and property is effectively protected. Basing assignments around the tactical priorities concentrates our individual and collective efforts and will bring a quicker (and safer) resolution to the incident problem. The standard priorities direct us in what we operationally do first, second, and third. Using the regular benchmarks of completion for the three tactical priorities ("all clear"/"under control"/"loss stopped") provide a standard set of signals in a predictable and understandable order, to indicate the current priority is complete and work can then begin on the next priority. This

Safety Effect:

regular work management approach (standard formation/standard plays) also helps the IC to effectively connect the standard risk management level that should go with each priority. This is where the "risk a lot to protect a savable life (rescue), risk a little in a highly calculated manner to protect savable property (offensive fire control), and don't take any risk to attempt to save lives/property already lost (defensive fire ops)" goes from a theory (blabbing) to an actual practice (doing). The standard priorities create the basic info management structure for the IC and produce a regular game plan that mobilizes where the troops go, why they go there, and what their objective is when they get there. This makes operations safer because they are predictable, dependable, and reproducible.

Command Safety

Situation Evaluation

IC Checklist:

- ☐ Evaluate current conditions and forecast future conditions along a standard scale.

Safety Effect:

The IC must always be thinking ahead of current conditions to prevent surprises (changes) that can hurt the workers. This proactive, surprise-prevention approach creates the capability for the IC to anticipate where incident conditions will go if the current action doesn't make the problem(s) get better and then go away. To forecast quickly and effectively, the IC must understand and use a regular scale that shows the standard stages that a structural fire naturally goes through (actually burns through), from nothing showing, to a smoking hole in the ground (i.e., total loss). The scale shows a series of "snapshots" that show how the fire progresses through standard stages, going from zero to ten. Below the one-to-ten snapshots, another one-to-ten scale is used to show the standard operational response that goes with each fire stage. The operational response scale starts with offensively investigating a "nothing showing stage one" to safely and defensively standing back (way back) admiring the magic of combustion in a fully involved stage nine/ten. Using the two scales (condition/response) together (parallel to each other) creates

Safety Effect:

the simple, practical capability for the IC and the ops team to effectively and safely connect the appropriate action that goes with that condition. The IC must (in effect) take a quick snapshot of the current stage, put it up to the one-to-ten scale, get a number of where the event is right now, use the standard action scale below the numeric incident stage to then connect the appropriate action that goes with that number. Using the snapshots that are beyond the current stage as a framework to describe "what's next" (and next and next, etc.) creates the very practical forecasting capability to understand what the future will look like if the problem isn't stopped, and what our standard safety-based response (position/function) should be in relation to those changes. Simply, using the forecasting scale, along with the standard risk management plan to predict the future when the IC decides the safety math is going negative (incident hazard is out performing the safety system), and moving the troops earlier rather than later, when incident conditions get worse (bigger numbers), and produce lots of pain prevention. The standard risk-management scale becomes the foundation for the very simple and practical model that connects the standard action we take to the conditions we encounter (standardized by matching the current stage to a place on the scale), and the standard outcome that we predict will emerge out of that standard action.

Command Safety

Situation Evaluation

Safety Effect:

This proactive safety approach creates the capability (and becomes the very practical foundation) for the IC to do everything possible to act out (in the real world) the standard conditions/standard action/standard outcome routine. This creates an effective level of tactical reality therapy for the entire team. No matter how much we would like to live in a world where we put out every fire offensively, that heavenly system just ain't in the cards, simply because we must play with a well-shuffled deck that has offensive, marginal, and defensive cards.

The full range of standard outcomes (from + to -) is the predictable result of the manual fire suppression component of the layer overall fire suppression system. That system starts out on one end with a lot of proactive responses: standard design and construction, built-in protection, fire code enforcement, etc. It ends up on the other end with a highly reactive response: manual fire suppression. The basic reason we must now respond and physically fight a fire is because some earlier part of the fire protection system has failed. Manual fire suppression is the most dangerous and least effective part of the overall system because there is now a good chance that something or somebody can get beat up, or worse, before we can effectively intervene. The IC must understand and effectively react to how operating on the reactive end of the scale actually occurs. All we can do is connect what we do to the conditions we inherit (from the failure), and

Safety Effect:

what will occur out of this connection is a standard outcome. We will get beat up a lot until and unless the IC understands this reality, and then effectively creates the appropriate overall strategy and IAP to match those conditions. All of this command and action must fit into the standard scale of conditions.

Developing the understanding and skill in actually connecting the "action" scale to the "condition" scale becomes the very practical basis of making consistent sense out of how "outcome" occurs. As we apply and connect the condition/action/outcome pieces over time, we become a lot more familiar and realistic (and conversant) about what the application of operational action can and cannot achieve. As we continue to apply this approach, we begin to standardize the relationship of the three "standard" pieces of the process... and this practice produces the recurring blab: "we must apply standard action to standard conditions to achieve a standard outcome." We don't say "positive, nice, desirable, favorable, pleasant" outcomes, we say "standard." If we ask an experienced street medic, "what's the standard outcome of being shot directly in the "business" part (as opposed to the cosmetic) part of the head with a large caliber bullet." They will calmly tell us: shot in the head (standard condition) causes us to apply advanced life support (stabilization), the standard outcome (of the condition, not the action) is dead right there (DRT). The medic cannot "unshoot" the customer, any

Safety Effect:

more than the IC can "unburn" the building. A major problem occurs when the IC operates as if applying offensive action to a defensive situation will somehow achieve some offensive miracle. When we encounter defensive conditions, no matter if they were waiting for us when we arrived, or if we screwed up what was a "temporary" offensive situation that just devolved into a "permanent" defensive situation, that defensive fire area is going to burn down... and it is going to do that with us in the hazard zone, or without us in the hazard zone. The (old) author has been looking at fire outcome pictures for almost fifty years. The defensive ones, ones that killed a firefighter, look exactly like the defensive ones that didn't kill a firefighter. A major (and very sad) part of this reality is that the standard outcome of having firefighters in offensive positions, under defensive conditions, is that they get beat up and killed... this isn't nice, but it's standard.

Command Safety

Situation Evaluation

IC Checklist:

☐ Continually reconsider conditions.

Safety Effect:

The IC expands and improves the overall situation evaluation capability by developing an incident organization that provides an ongoing stream of information on the status of incident conditions all over the incident site. The IC "connects" to the entire situation by assigning companies and sectors to all the critical incident areas and functions. Creating a regular approach to this evaluation and reporting game plan ahead of the event provides a huge organizational effectiveness head start for the entire team and puts everyone on the "same page." The IC combines this decentralized reported info (i.e., info "push" from deployed units to IC) with what can be seen from the CP to create and maintain a balanced ongoing evaluation (visual/recon) of the entire incident situation. Maintaining an awareness of what is currently going on, in and around the hot zone, requires the IC to intensely process information from every critical place and source. The IC must also assemble, integrate, and process info coming into the CP from every one of the critical spots. Sometimes one reporting position will say, as an example, that the fire is under control (which it is there) and another place will be calling for more water because the fire is kickin' their ass--or the inside firefighter declares fire control victory and the IC can see an eighty-five foot flame front boomin' out of the roof (a very obvious "red

Safety Effect:

flag"). The standard scale (presented in the last section) provides the framework that this incoming info is continually integrated into and compared against. Incident action either overpowers the problem, and things get better--or the problem overpowers and gets ahead of what we are doing, and things get worse (stalemates are "red flags"--all ties go to the fire/gravity). It is a challenging (and absolutely necessary) process for the IC to quickly develop the on-line capability to effectively receive and process information from every basic operational/functional position about the critical incident factors that affect worker safety. This is particularly important when there are workers operating in the hazard zone and the situation is not stabilized. These are situations where the incident problem may be able to outrun the firefighters to the exit(s). A major fatal firefighter situation involves what appears to be a highly manageable fire situation that firefighters just naturally move in on and get close to. Many times they are in a hurry(!) so they either don't initially reinforce and "harden" the way they came in (exit paths), or don't order those resources (multiple hose lines, RIC teams, lights, guide/usher companies, safety officers, etc.) in behind them. If the fire changes quickly (bigger/hotter/badder), they can become trapped and now the situation goes from an underwhelming routine fire fight to an overwhelming desperate rescue. We must build and manage the rescue capability before anybody gets in trouble. These conditions present big-time safety risks, so the IC had better be awake,

Command Safety

Situation Evaluation

Safety Effect:

paying attention, and ready to call quick safety plays, based on incoming info and effective forecasting. The IC must always be thinking, planning, and getting ready to add plan B to the end (or, sometimes the middle) of plan A and be mentally and operationally setting up to first prevent what causes bad news--but also thinking about the when/where/what that will be required to effectively react to bad news. This is the essence of our continual blab about us doing the standard routine five-hundred times where it's not needed which automatically places us in the correct (safe) place, the one desperate time when we do need it. Like all life preservers, when we need it, we always need it badly (and we survive based in its availability and our training and preparation). The IC lives in a dream world if they think they can just sit in their air-conditioned suburban, look regal, munch a Snickers bar, and have a steady stream of top-drawer information automatically and naturally land in their lap. Effective incident info management requires the IC to both create an incident organization by making assignments in critical places, and then aggressively working that organization ("pulling info") to continually get the right information from the right

Safety Effect:

spot at the right time to produce the safety math that protects the troops--simply, the IC had better be as "hot" as the hot zone.

Command Safety

Situation Evaluation

IC Checklist:

☐ Interrogate/interact with the owner, occupant, building engineer, technicians, etc.

Safety Effect:

In large, complicated, technical (and many times hazardous) places like large, complex multiple occupancy, industrial, commercial, high rise, manufacturing, storage, refinery occupancies, it is helpful and smart for the IC to develop an up-front (i.e., early in the incident) liaison with a responsible party who has an intimate knowledge and understanding of occupancy details. The more complicated the place, process or situation, the more important it becomes for the IC to assign this role to a person dedicated to the liaison function who is part of the command staff. We use the word "local" a lot to describe a hometown response... local simply doesn't get any more local than dealing with the occupancy rep. Simply, the occupancy reps live there, work there, sometimes have built the place, and are the folks who operate it--they know the most about it. It is impossible for us, under typical incident conditions (or really any time), to gather, understand, and use their level of local familiarity while the event is underway. Many times, the occupancy rep (usually the person with the most keys on their belt) can tell us about the very recent history and status (hopefully, instant recall) of whatever is causing the current problem (lots of times they are the ones

Safety Effect:

who called us). They may have useful ideas and experience in how to use internal, building-process systems to assist us in solving the incident problem. Creating a technical sector (or a planning section in bigger events) to deal with the occupancy rep is a useful organizational approach. This sector can take the time to directly interact with the occupancy rep so they gather info effectively, and then translate and forward that intelligence on to the command team. Selecting a fire department member with special skill/training/certification (haz mat, special ops, fire prevention, fire protection engineer, BC, etc.) who speaks occupancy lingo is smart. Technical-liaison sector officers should also have good listening skills and must consistently demonstrate good playground behaviors (play well with others).

Command Safety

Situation Evaluation

IC Checklist:

☐ Maintain a realistic awareness of elapsed incident time.

Safety Effect:

Most local incident ops are conducted very quickly within a compressed time frame. Many times these rushed conditions create a serious time distortion for the IC. Simply, a lot of times when the IC has been operating for twelve minutes, it feels like two minutes. It requires skill, experience, and system support for the IC to effectively keep track of elapsed incident time. Being able to manage time becomes a critical command activity because the hazard zone has an effect on anyone and anything that is in it, and that effect is directly connected to the duration of their presence in the hazard zone. Examples of the effect of time are the humans who are trapped and require rescue are many times holding their breath; the structure typically can only survive an uncontrolled fire for a short length of time; the air in firefighters' bottles is finite. The IC, generally, has a narrow (and many times rapidly shrinking) window of opportunity to make a positive offensive impact on the critical factors at an active structure fire. The IC must develop a "clock ticking inside their head" approach that connects conditions/action/outcome with time. The IC must forecast conditions and outcomes based on the time it will take to do basic operations... things like primary search operations and to place and operate attack lines. Everyone must understand initial operational setup

Command Safety

Situation Evaluation

Safety Effect:

times and be aware that the fire is generally growing (to the third power), and the fire you find (when you get into that attack position) will be bigger than it was when you started pullin' hose. This is why successful hunters shoot ahead of flying ducks and let the ducks commit suicide by flying into the buckshot. If the IC doesn't understand this, they will always be shooting behind the ducks (fire). It is always critical that the IC also consider the time it takes for inside crews to exit. A time-conscious IC must estimate the time the fire has been burning before their arrival, and then must factor that time into the remaining offensive time. The IC must develop the skill to make such an evaluation of how long the fire burned before operational action occurred, by applying previous experience to the current incident. ICs must pay attention to the relationship between fire conditions, structural reactions, and time. These time estimates are not absolutely accurate, but are the best chance the system has to safely connect the overall strategy to total elapsed incident time. We must not be distracted by thinking that the event started when we arrived--it actually started when the fire arrived. It can be a huge error to think we are in the beginning stages of the event when, in fact, we are actually in the end of the middle, where changes are typically picking up speed and moving rapidly toward the beginning of the end. When the IC determines the overall offensive time (window of opportunity) is twelve minutes (as an example), and estimates the fire has been burning eight minutes, they must

Safety Effect:

then indicate to insiders, "You have four minutes to improve conditions"--the IC must then set the egg timer on four and when it goes "bing," if the problem hasn't gotten better, the IC must tell the attackers to say "adios" to the inside, and move to a safe place. TWSs must have a checkoff time line that corresponds to standard ten-minute ET reports that are transmitted from the commo center. The IC must specifically acknowledge ETs (i.e., repeat the time) and must reconfirm the overall strategy. The IC must also obtain a PAR at standard time intervals that verifies the welfare of all members in the hazard zone.

Major safety problems occur (many times sadly) when there are products of combustion present (somewhere... ?) and inside troops can't find the fire. Then, suddenly, the fire finds, surprises, and then overpowers (injures or kills) the exposed workers. In these potential "sucker-punch" deals, the IC must micromanage having evidence of a fire being present (smoke/heat) and not knowing where the fire is (or is going), as a major unknown critical factor. The IC must evaluate how much time the troops have to keep looking before the structure falls down, lights up, or traps them. The IC must mark that time, set the "timer" and when it goes "bing," if the problem isn't accurately identified, evaluated, and effectively responded to, or if the known problem isn't getting better, get 'em out. Effective time management becomes a major part of this safety process. It's tough to pull the troops out before they have found the problem because they are (thankfully) highly

Safety Effect:

motivated to intensely and aggressively engage incident conditions so they can solve the incident problem. It's lots better to have a bunch of angry firefighters watch a building collapse while they are having a Snickers experience in rehab, than it is to have a collapse experience in the hot zone because of an IC time-management breakdown, in which losing track of time is a normal (and predictable) response to the degree of IC stress that is typically present during fireground operations. Managing such time-distorting stress requires self discipline, job aids, and closely connected friends (i.e., team management) to overcome.

A major part of "reading" the gauges presented earlier in this chapter involves paying attention to the time factor of the changes that are occurring with that particular condition. How fast that condition is changing (either bigger or smaller) becomes a major way the IC evaluates current and future conditions. Very little is static on the fireground, so the IC must develop effective time-management skills to protect the troops.

Command Safety

Situation Evaluation

IC Checklist:

☐ Consider fixed factors and manage variable factors.

Safety Effect:

Dividing conditions into fixed and variable factors creates an interesting management challenge (and opportunity) that requires a highly practical and realistic response from the IC. It involves the IC developing the ability to sort out and separate what we can control, from what we have very little ability to influence. This requires the IC to consider all the fixed and variable factors (in both categories: able to control/not able to control), and then to focus on going to work on the ones that the response effort can do something about. This approach eliminates a lot of wasted command and operational effort, and makes it easier for the IC to pay attention to fixed conditions that cannot be changed by our action, but that can be highly dangerous to our workers, and that absolutely require the IC's attention. As examples, if the industrial process involves high explosives or if the commercial building is so big they can land airplanes on the roof, there isn't much the IC can do (at incident time) to change those realities. The IC cannot cause the building to shrink (the fire can) or to "turn off" gravity (imagine firefighters floating in the air, only tethered to the earth by their

Safety Effect:

hose lines... nomex balloons). What the IC must do is establish an overall strategy, IAP, and safety plan that realistically considers those conditions.

Fixed factors are the things the IC can't change--like the original size, arrangement, age, and construction of the building. The presence or absence of fixed fire protection systems and water supplies are also among fixed factors. These are the physical factors that lend themselves to being visited, studied, and recorded earlier on a pre-incident plan. Knowing about these built-in physical factors, increases the IC's capability to develop a safe and effective IAP. Variable factors relate mostly to the status of what is causing the incident problem--like the location and condition of threatened humans and what it will take to protect and rescue them. The IC must engage a quick, dynamic evaluation and information management system to develop an awareness of the details of the variable factors. The location and extent of the fire is always a critical (and we hope controllable) variable factor--if the IC can extinguish the fire, everything will get better. Fires do not make buildings stronger, so structural integrity is a critical variable factor that will continue to worsen until the fire goes out (either by our suppression efforts or by running out of fuel) and/or the building collapses. The IC must concentrate on gathering and processing initial and ongoing information on the critical factors that are out of control, and developing an IAP that protects the incident workers while they bring those

Command Safety

Situation Evaluation

This happens when Situation Evaluation is not done:

- ☐ Lack of good info produces incorrect placement and action.
- ☐ No forecasting... difficult to anticipate and react to changing conditions.
- ☐ IC fails to develop critical-factor inventory for the incident--has no effective knowledge of knowns/ unknowns.
- ☐ Continual panic is triggered by lots of dangerous surprises (reported after they occur).
- ☐ No good information causes IC to sound like an idiot.
- ☐ It is difficult to develop and maintain an effective action plan.
- ☐ No effective info management produces fragmented reporting and disconnected action--IC is not aware of what info is needed (or by whom).
- ☐ IC misses and must continually react to "red flags."
- ☐ Troops get mad when IC does not react to quality (or really any) information/condition/action reports.
- ☐ Lots of unsafe free-lancing.
- ☐ Little chance of ever catching up.

Command Safety

Situation Evaluation

COACHING VERSION

"Avoid fancy evaluation systems during fast-and-dirty, initial fireground stages. Look at what's going on from where you are, pick out the critical stuff, forecast where the problem is going, get someone (company/ sector) on the interior and opposite side from you who can report on inside and side C conditions to eliminate surprises. Try to convert assumptions to facts--don't be distracted by minor stuff--don't let the fire hypnotize you--avoid tunnel vision--stay awake and ahead of surprises. Use the standard inventory of critical factors that go with that situation as a routine checklist to record what you know and don't know. Aggressively and quickly go after the critical unknown factors. Pay particular attention to safety "red flags"--virtually every "red flag" gets more severe if we allow it to get older... don't hope they will go away by themselves--when you see "red flags," quickly react to them... conditions either get better or they get worse--there is very little actual "holding" of fire conditions. Simply, this is why "shit happens." You must either eliminate the problem or get out of it's way. Don't let a lot going on in the beginning overload you. Sort out and line up information around rescue/fire control/property conservation, and focus on going to work in that "one-two-three" priority order (sometimes you must do "two" [fire control] in order to safely do "one" [rescue]). You must retain your command stature and sanity. The fire will do everything it can to distract you, particularly in the beginning of the event. These distractions can make effective information management impossible. Trust your gut--if it tells you it's bad--react and then continue to refine that reaction. If looking at the fire makes you nuts, turn your back on it, and concentrate on doing the command functions. "This is a learned skill--keep practicing/learning/doing."

Command Safety

Situation Evaluation

Command Safety

Communications

CHAPTER 3

Command Safety

Communications

Command Safety

Communications

Function 3

Communications

Major Goal

To initiate, maintain, and control effective incident communications.

IC Checklist:

- ☐ Use communications SOPs.
- ☐ Start/control communications upon arrival with IRR that describes conditions and actions.
- ☐ Use effective communications techniques to keep everyone connected.
- ☐ Use organizational chart as communications flow plan.
- ☐ Use company, sectors, and communications center as communications partners.
- ☐ Maintain a clear, controlled, well-timed radio voice; project a good radio image.
- ☐ Listen critically--understand communications difficulties from tough operating positions.
- ☐ Mix and match forms of communications (face to face--radio--computers--SOPs).
- ☐ Coordinate timely progress reports.
- ☐ Maintain your communication availability.
- ☐ Utilize the standard order model to structure communications.
- ☐ Keep communications simple: task/location/objective (use plain text).
- ☐ Utilize CP position and staff to improve communications.
- ☐ Center communications around the tactical benchmarks--"all clear," "under control," "loss stopped."

Command Safety

Communications

IC Checklist:

- ☐ Use communications SOPs.

Safety Effect:

Creating a standard way the response team will use to commo with each other becomes a major factor in the safety process. IMS uses standard commo to get the team into the hot zone, to evaluate their safety and effectiveness while they are there, and uses it to quickly get them out if necessary. An ongoing and effective commo connection between the inside (hazard zone) and outside (command/support/organization/CP) is a critical component of offensive operations. The lack of this ongoing inside/outside commo link is always a major "red flag." The IC must continually monitor this inside/outside connection whenever there are firefighters working in the hazard zone, and if that commo connection is broken (for whatever reason), the IC must either fix the commo or change the strategy. It is simply impossible for the IC to make such changes (that control the position and function of the haz-zone troops) without effective commo, because such effective commo is the way the IC causes the haz zoners to move out of harm's way. Commo that works is a major part of what makes safety work (... and work safe). Commo is so critical that it is now a common fire service SOP that if a haz-zone team cannot maintain continuous

Command Safety

Communications

Safety Effect:

commo with their IC/sector, they are essentially disconnected from the safety system (command team, RIC, etc.), must exit and either reestablish commo or stay out. Communications SOPs provide all the incident players with a common language and a regular way (i.e., commo system) to use that language to exchange information. That language is "spoken" (and exchanged) in a tough setting. Typical incident conditions are noisy, dark, rushed, and potentially or actually dangerous. Commo participants are in protective gear, so they are many times shouting through SCBA face pieces into portable radios. Everybody in the hazard zone is screaming, grunting, shoving, and moving quickly. These lousy incident conditions make it particularly important for us to develop a practical, doable commo routine (described in local SOPs) that we can agree on, learn, practice, and refine our techniques and SOPs ahead of having to go into the commo hot-zone environment and somehow attempt to stay connected with the outside (IC in CP) world under pretty lousy commo conditions. A major part of that commo-based approach is for the IC to be (as quickly as possible) "locked up" sitting inside a CP located close to, but outside, the hazard zone in a vantage point (good view) position where they can use the advantage of being in a good commo spot to help the troops (on the inside) who are in lousy commo spots.

Command Safety

Communications

IC Checklist:

☐ Start/control communications upon arrival with an IRR that describes conditions and actions.

Safety Effect:

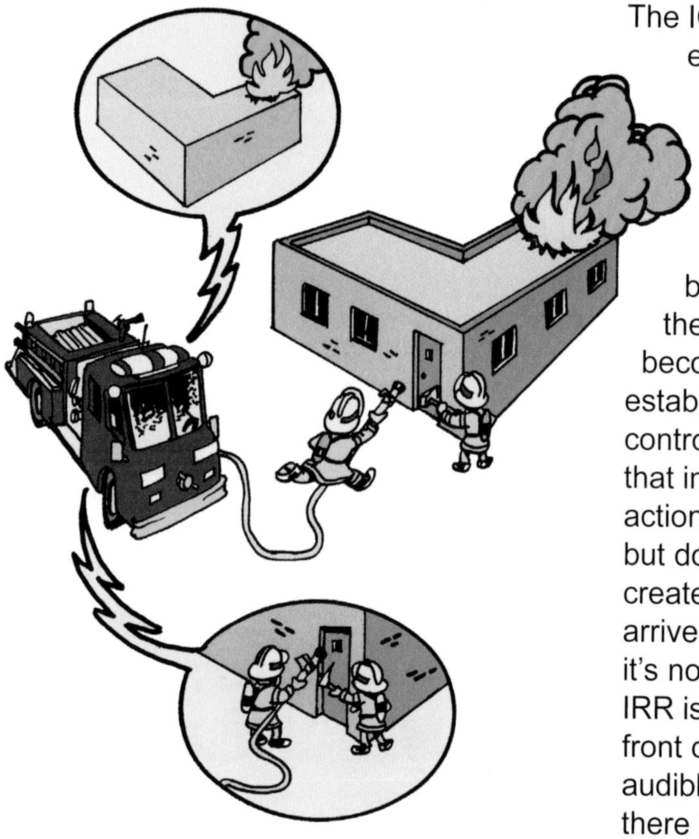

The IC sets up in the very beginning of the event as the incident communications sheriff with a good on-scene report that specifically includes the assumption of command. The IRR is a simple and standard way for the IC to take control of the commo process by being the first responder to talk on the radio. While this is pretty simple, it becomes critical that we use the IRR to establish early management and operational control. If there is a lot of radio blab before that initial report is given (generally by action-oriented initial arrivers who should, but don't assume command), we start to create the beginning of a cluster. Whoever arrives first gives the initial report (or else it's not initial) and becomes IC #1. After that IRR is given, IC #1 has established an up-front commo "anchor point." Now there is audible evidence that an IC in place, so there is an IC in place to commo with.

In the absence of an effective IRR, two things typically occur:

1. No one says a word on the radio--the first arriver skips taking command and just goes to work (free-lances)... or

Safety Effect:

2. First arriver screams non-command blab ("Martians landed") into the radio just before they blast off to actionville.

Both one and two provide a perfect front end (launching pad) for an unsafe, uncommanded cluster. The more responders who arrive and don't establish commo and command, the bigger the mess becomes, and the more difficult it will be for some response boss to eventually take (wrestle) command, establish effective incident control, and start to round up the free-lancers and get them into the IC's plan.

The IRR requires IC #1 to describe the incident problem and what basic actions are being taken to solve those problem(s). This front-end information connects all the incident players. This standard initial report puts everyone on the same page and starts incident operations with the whole team being part of the IC's game plan.

Before any fire company (or anyone else) goes into the hazard zone, three things should be effectively in place:

1. There is an IC set up and operating.

2. The company must be intact and wearing the proper PPE.

Command Safety

Communications

Safety Effect:

3. They must have a specific assignment (either by order, SOP, or conscious officer decision reported back to IC) and they have an assigned boss.

If those three basic conditions are not present, DO NOT GO INTO THE HAZARD ZONE. The major way that subsequent-arriving responders can tell if there is an IC in place and operating is when (and if) that IC gives an IRR (that a sober person can make sense of).

A regular part of the IRR that is critical to incident safety is the standard assumption and confirmation of command. Having the first arriver make a conscious and deliberate "public" expression that they will be the IC: "E-1 will be Main Street Command," and then having the communications center repeat (parrot) the confirmation back over the tactical channel is a big deal. It insures (and announces) that a real live person is actually on the scene and is (at least) lucid, oriented and rational enough to say, "I am in command," and for the commo home office to repeat it back for all the world to hear. Real simple: responder says it = assumption/ commo center repeats it = confirmation... now we can go to work with an IC in place. This simple but absolutely critical procedure is how we give birth to IC #1 in a standard way. This creates a real simple, short, and sweet (standard) incident beginning, and gives us a fighting chance of developing a plan to go to work on the incident problem,

Safety Effect:

and to both get into and then get out of the hazard zone (i.e., "round trip"). Everyone operating at the incident must somehow connect to this assumption/confirmation occurring as a standard practice. Where/ when it does not occur, the commo center and response bosses must "encourage" the first arriver to become the IC. We must realize that whenever we screw up the standard, front end of command (assumption and confirmation), we are in a potentially very dangerous situation... there is a hazard zone present, units have arrived on the scene, no one has assumed command, and it appears that firefighters have or are about to enter the hazard zone. This is why the system must be very anal retentive about always having a conductor in place before we break out our instruments and start playing.

Command Safety

Communications

IC Checklist:

☐ Use effective commo techniques to keep everyone connected.

Safety Effect:

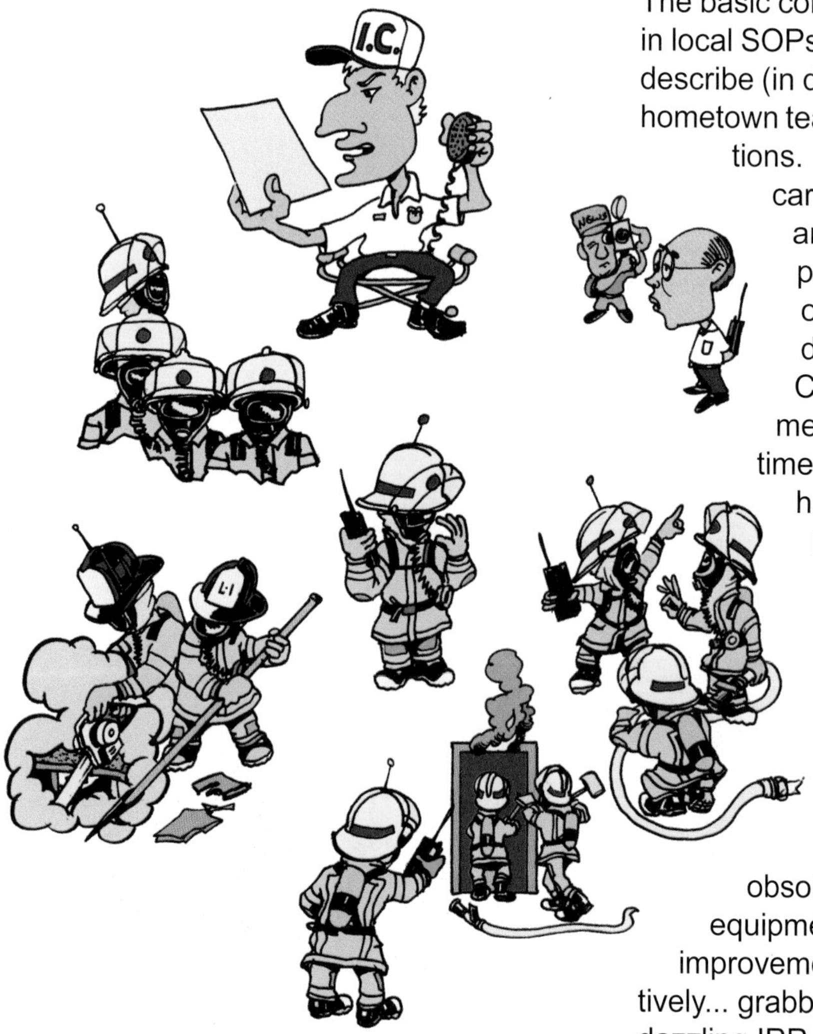

The basic commo game plan must be outlined in local SOPs. The SOPs are designed to describe (in detail) the basic game plan the hometown team will use during incident operations. How those commo SOPs are carried out and how effective they are is highly dependent on the personal techniques the IC uses to operationally apply those SOPs during the command process. Current electronic commo equipment is sophisticated and sometimes pretty complicated. We now have multiple radios, cell phones, mobile data terminals (MDTs), fax machines, personal computers, palm pilots, and who knows what else in the cabs of response vehicles. Most of these rigs do not come equipped with your fifteen-year-old grandson, Alex, to explain all the blinking lights to an old, electronically obsolete IC (grandfather). While this equipment creates significant commo improvements, it must be operated effectively... grabbing the wrong mic and giving a dazzling IRR over the outside PA system may be a thrilling experience for the spectators, but it doesn't do much for the rest of the response team (and the commo center) who is wondering what's going on out at the scene. The IC must develop functional voice control,

Safety Effect:

choice of words, and thoughtful commo timing techniques to stay continuously and effectively connected. The application of these functional habits, along with using the CP advantage, creates the capability for the IC to become the commo focal point for maintaining control, and for providing information support to the entire operation. The IC uses and relies on the entire on-line commo process more than any other person or part of the IMS--creating and maintaining strategic level control of the incident is about the IC always being able to commo with the tactical and task levels. The IC must maintain commo contact with every part of the organization at the scene and with the outside world. Through the commo center, the IC establishes and maintains the standard commo process mostly to use those commo resources to support and protect the hazard-zone workers. Basically, the IC communicates and acts out their strategic safety responsibility over the radio, and they can't do it if they are saying the wrong thing, at the wrong time, the wrong way, on the wrong channel. How the team exercises control over their part of the operation becomes a major safety factor. Companies must follow operational and safety SOPs--sectors must directly supervise their teams and look out for their welfare. The IC does their part to provide overall incident management by doing the standard functions of command in the CP. How the IC uses the commo system to stay connected to the hazard zoners will determine the IC's effectiveness (and the safety of the workers).... the workers use their tools to do their manual labor/

Command Safety

Communications

Safety Effect:

the IC does a major portion of their work with a radio mic, which becomes a major IC tool. Effective and safe fire fighting operations absolutely require a calm, lucid IC who can communicate in a disciplined and controlled manner, particularly during difficult situations. This capability becomes a reflection of the very personal skill and capability of the individual who is serving as the IC.

Note:

The authors remember in the interesting early days of command system development the reaction of our action-oriented firefighters to the new commo approach was "now we are talking fires out!" Before we developed (and then struggled to actually apply) commo SOPs, we mostly exchanged information by yelling at each other... so it was predictable that using the radio was regarded as pretty audibly bureaucratic. The early response was (and still is) accurate in that the IC mostly plays their strategic part by connecting to the outside world (commo center, sectors, companies) using the radio. Developing the capability for the IC to actually stay in a strategic CP position by using the commo function required creating an entire IMS (the other seven standard command functions) that effectively keeps everyone connected. In the early days, we never realized how important the commo process would become to responder safety and survival.

Command Safety

Communications

IC Checklist:

- ☐ Use the organizational chart as the commo flow plan.

Safety Effect:

Having the IC evaluate conditions and decide on what work needs to be done, and the support that work requires, is a major IMS objective. The IC then creates an incident organization to cover the critical geographical and functional parts of that work plan. A standard work cycle process is used throughout the operation that provides a framework for the regular way the system assigns and accounts for companies, as they go from responding to staging, and then going to work under the command of a sector officer or the IC. This basic incident organization creates both the template and the flow chart that is used for organizing and managing commo. This commo chart (which is a reflection of organization) becomes the very practical road map for who will commo with whom (the commo SOPs also describe the "how"). The IC uses the commo process to actually establish the incident organization. The SOPs describe how that local organizational development will

Command Safety

Communications

Safety Effect:

occur--then when there is an active incident going on, whoever says they are IC #1 over the tactical channel becomes the IC, and automatically begins to do the standard functions of command. This command announcement starts the process and all the ongoing commo continues to make the assignment (described in SOPs) that builds both the work process and the command structure... basically, it all occurs on the radio. Connecting the incident organization with the commo process becomes a critical incident safety factor. It creates a natural grouping of the bosses and workers assigned to all the major functions. This grouping makes it lots easier for those boss/worker "bunches" to commo back and forth... particularly between the IC and all the hot-zone sectors and companies. This standard commo connection directly links those who could get stuck or have critical needs in the hazard zone, with the person (sector/IC... "up the line") who is responsible for their ongoing welfare. The IC also must always be the person who is the most aware of their overall hot-zone status, and is the person who has the best access to the resources that can assist workers who need help--like RIC teams, accountability resources, safety officers, back-up crews, more water application, ventilation, etc. Typically the IC and sectors commo using mobile (IC) and portable (sectors) radios, companies operating within sectors will (as much as possible) commo face to face with sector officers. In long-term incidents, putting hot-zone support functions (PIO, safety,

Safety Effect:

rehab, water supply, etc.) on alternate radio frequencies protects and shelters (i.e., puts that traffic on another frequency) the tactical channel between the IC and the insiders. The IC and the command team must do whatever is required to maintain the integrity of the commo link between the hazard zone and the CP as the highest priority. Commo SOPs must be designed and directed to automatically support this haz-zone/CP connection.

Command Safety

Communications

IC Checklist:

☐ Use companies, sectors and the communications center as commo partners.

Safety Effect:

The IC sets up strategic command in the CP to become "information central" to quickly develop an incident organization that will recon and report the info required to develop a current picture of critical factors all over the incident back to the IC/command team. In order to do this, a decentralized evaluation and info exchange system using companies and sectors must quickly and (automatically) feed info ("push") into the CP. The IC then puts visual and reported info from all over the incident site together ("adds it up") to assemble an overall picture that is used to evaluate the ongoing safety and effectiveness of operating positions. Making it a standard and automatic part of every assignment to forward info about conditions/action in that area/function up the chain of command is what creates this capability. Having the IC send information that has been evaluated, processed, and integrated back down into the operating part of the organization reinforces the two-way commo process and creates the confidence within the operational team that the IC is in the CP and is awake and listening. This

Command Safety

Communications

Safety Effect:

ongoing, highly responsive commo process creates the confidence of the hazard-zone workers that the IC is paying attention and reacting to them when they punch the button and talk. The IC is in the best (sometimes only) position to collect and rationally interpret info from all over the incident, and then use all that intelligence to continually adjust the safety of the overall strategy and IAP. The IC is also, along with the communications center, in the best place to quickly distribute critical directions down into the organization that moves the troops away from danger (like going from offensive to defensive) and when this happens, then reverifying that the insiders have a PAR and are now operating in standard, outside defensive positions. Good commo is about both sending and receiving (talking/listening), and requires good partners on both ends of the process who understand the regular and special conditions that go along with the other partner's position, and how those conditions will influence their ability to commo. This understanding is particularly important on the IC's part... whenever a hazard-zone transmission comes into the CP, the IC must instinctively connect the typical conditions that exist where that sender is operating. Hazard zones are lousy places to try to communicate--noisy, hot, rushed, saws grinding, fans running, radios blaring, people yelling. The IC must provide patience, support, and critical listening to make that commo work. This becomes one of the many places where "respect the task" fits.

Command Safety

Communications

IC Checklist:

☐ Maintain a clear, controlled, well-timed radio voice image.

Safety Effect:

Radio is the major tool the IC uses to interact with everyone outside the CP, so the IC's presence is mostly conveyed over the radio... yes, Virginia, the IC does "talk the fire out!" If the IC speaks with a clear, calm, and rational voice, people will be more inclined to listen to what they have to say. How the IC times their talk and integrates it into what others are saying sends a powerful message on how the IC is paying attention, listening, and connecting ("relating") with others. Screaming lunatics do not instill much confidence in the work force (however, they do make for great firehouse jokes). Also, an IC who is audibly and obviously out of control also does not register very high on the listening and reacting scale, and doesn't create a very positive feeling that they have the basic sanity to hear and react to a response that is sent back to them. The IC's radio demeanor is infectious. If the IC is out of control over the radio, the workers may very well follow suit. How the IC sounds on the radio sends a strong message on the level of self control the IC has over the IC (duh!). Operating sectors/companies lots of times are huffing, puffing, and yelling into a portable radio--how they sound is a very legitimate reflection of where they are and what

Safety Effect:

they are doing... working. It is reassuring to such hazard-zone workers to have a lucid, calm voice come out of the confusion, and supportively (and effectively) react over the radio to the worker's sometimes not very calm commo. The calm voice in the chaos builds confidence that the IC has their head out where the sun is shining and is looking after the safety, welfare, and survival of workers who can be working in pretty gritty spots (where the sun ain't shining). Creating the consistent ability to maintain control of the incident organization and operation must begin with the IC. A major way that level of control is evaluated (and then reacted to) is simply by listening to how under control the IC sounds on the radio. Just about every part of performing the IC's command and safety functions involves talking into a radio mic, so it is worth whatever organizational effort is required to create a high level of command commo capability.

There are just naturally a lot of people, places, and things present at an incident that can get under the IC's "skin"--a lot of these incident players cannot receive, react, or sometimes even understand what the IC is trying to do. Sometimes (actually a lot of times) these confusing incident elements can piss off the IC. This is a bad thing because becoming angry will always reduce the IC's effectiveness. A major place where this occurs is how being mad will affect how the IC sounds when they communicate. The answer is difficult (to

Safety Effect:

actually and consistently do) but is absolutely essential--DON'T LET ANYTHING OR ANYBODY MAKE YOU MAD... or, maybe a better way to say it, is don't let anybody (else) know you are mad when you talk on the radio--the deodorant commercial says it best, "Don't ever let 'em see you sweat." When you lose your head, the next thing's your ass.

Note:

"Don't get mad over the radio" can be a big (safety) deal. The strongest message is in the tone of the IC's voice. Hazard-zone workers have done some screwy stuff (sometimes fatal) because they received the message that their boss was in some way upset, mad, or acted out of balance with them. When the IC wants a hazard zoner to move quickly, it's okay to deliver a direct order with enough emphasis ("get off the roof!") that the order sounds urgent, but don't make it sound sarcastic. How the boss sounds is always important to the worker.

Command Safety

Communications

IC Checklist:

☐ Listen critically--understand commo difficulties from tough operating positions.

Safety Effect:

The IMS creates the task, tactical, and strategic levels of operation--these separate, but closely connected levels, all operate with different commo capability. The task level involves fire companies doing manual labor in the hazard zone. Their typical operating position is right where the work is being done (by them) so it is many times noisy, wet, dirty, exciting, and dangerous. It's a lousy commo position. Sector bosses work on the tactical level and directly supervise fire companies. They are necessarily close to where the work is being done (simply because they are supervising the work teams), so they are also in disadvantaged commo spots. Sectors and companies commo with each other mostly face to face (i.e., yelling/grunting/hand signals). The IC and sectors talk to each other on the radio (sectors/portables--IC/mobile). A major way the IMS compensates for the disadvantaged (and lots of times hazardous) commo positions of the companies and sectors is by placing the IC in a good commo spot--the CP. Being inside the CP (remember the caveman deal) is as close as we can produce to an office-like environment (even though it's pretty crude). This creates the best on-scene capability to critically listen to

Command Safety

Communications

Safety Effect:

incoming commo coming from the hot zone that is many times not too "clean"--simply, when someone keys a mic in the haz zone, it's pretty easy to hear fire streams, saws, stuff falling, people yelling, "B" shifters grunting, etc. The IC must critically "listen through" such environmental noise for the message, acknowledge that message (once it is understood), and quickly react to it. The IC has a fighting chance to do that sitting INSIDE a quiet CP with a head set, a current TWS, and some command-team helpers. It is absolutely impossible for the IC to effectively hear and react to such commo standing outside, surrounded by yelling players jumping up and down, goosing each other(!) so they can win the competition for the IC's attention.

A huge safety problem is created when the IC routinely operates in a disadvantaged command position and for an extended period gets away with outside (the CP), aerobic, roving, face-to-face (yelling) behavior. Aggressive, skillful, task-level performance will solve the incident problem most of the time--in fact, creating and managing such problem solving performance is the basic objective of the IMS. But the IMS requires an effective strategic level of command support to always be in place to establish, reinforce, and protect such effective, task-level performance. We really don't know if we have that operational support until something unusual, surprising, or bad happens and we need to quickly (right now!) and

Safety Effect:

effectively (get it done) react to a safety problem that is present in the hazard zone. Having an effective, awake IC inside a CP located close to but outside the hazard zone (looking, listening, receiving, thinking, deciding, reacting) is like wearing your seat belt. Even though you don't need it most of the time, after awhile, you just put it on automatically, so wearing it just becomes a habit. This becomes a very good thing because you had better be wearing it when you need it, because you don't have the time to put it on when your head is rocketing through the windshield. Keep it simple: always wear your seat belt; always take command. They both save us.

Command Safety

Communications

IC Checklist:

☐ Mix and match forms of commo (face to face, radio, computers, SOPs).

Safety Effect:

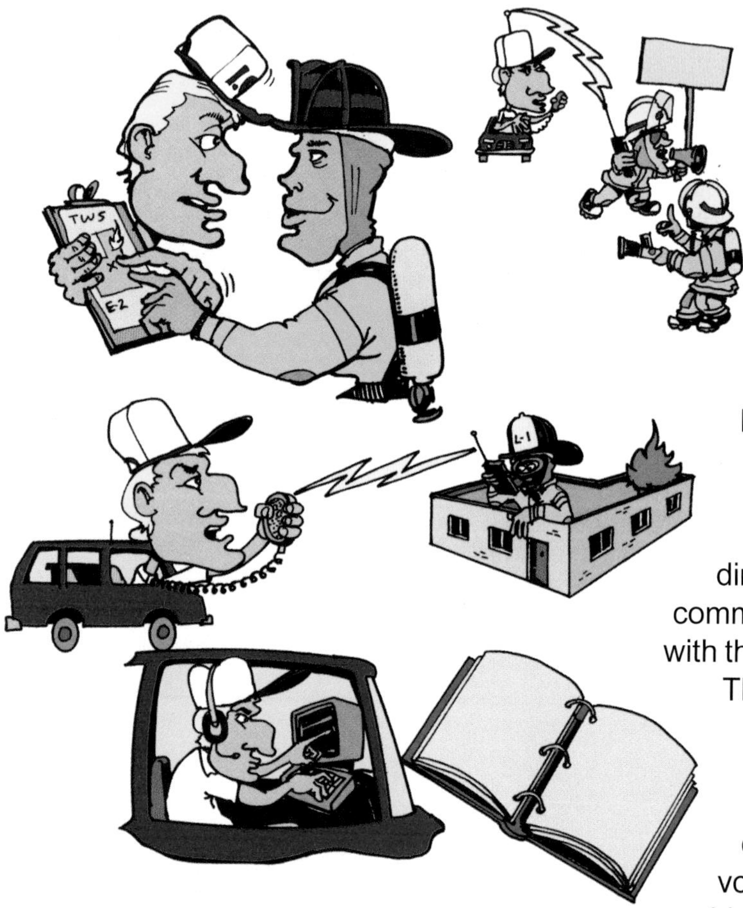

The IC must take a smorgasbord (pick and choose) approach to using the different commo forms. Each form has it's own set of characteristics. Face to face works best where the participants are close together (physically) like between the command team inside the CP and between working companies and sector officers. It is direct, short range, and the most personal form of commo because it is both verbal (talk/hear) and nonverbal, up close and personal (talk/look/respond). It is necessarily restricted to those who can see and hear each other personally and directly like when the IC assumes the command mode position and commos with the command team (inside the CP). The radio is how the IC gets their voice (and message) out of the CP. The radio is wide ranging, quick and one dimensional (voice only). The level of confidence in the voice of the commo players becomes a big part of the "message." Radio commo is very fragile and is easy to screw up if the participants don't play nicely. Whoever pushes the mic button captures that radio frequency, so if everyone doesn't follow

Command Safety

Communications

Safety Effect:

the commo SOPs, we essentially lose that frequency. The IC must always use the other commo forms in a way that compliments, supports, and clears the way to maintain a strong radio link to the safety of the hazard-zone workers. When the IC must move the workers quickly, there isn't time to send a detailed internet message or a singing telegram. Simply, "abandon the building!" over the radio (i.e., tactical channel) works a lot better. The IC must always make sure this critical type of message gets complete acknowledgment from all sectors/companies that are in potential danger. Computers are a common way to store and quickly retrieve a lot of info. They are fast, can maintain lots of data, are highly dependent on good software, current information, and a capable operator. They work best on information that can be loaded in ahead of time. SOPs create the capability for the organization to decide on how they will conduct commo activities (along with a standard language) in standard situations. When they encounter such regular, recurring conditions, the SOP creates a quick way to trigger a standard response. This SOP approach uses a small number of words to create a big (automatic) response, which is a good way to produce standard (effective and safe) organizational and operational reactions.

Making this planned, practiced, information management approach a part of our regular command and operational routine creates a

Command Safety

Communications

Safety Effect:

major safety capability. When the IC must quickly react to changing conditions, there is no time for elaborate, long or involved commo--then it is show time for the command safety system and everybody had better be on the right channel, talking, and listening in "safety shorthand." This time and place (when we must quickly react) becomes the "moment of truth" for our command and safety system, and how well it works (or doesn't work) becomes a reflection of how we have developed the system and routinely operated in the past. What we have done routinely on the events before this one will have produced the functional (safe) habits or the dysfunctional (unsafe) ones that will either save us or kill us.

Command Safety

Communications

IC Checklist:

- ☐ Coordinate timely progress reports.

Safety Effect:

Progress reports become a major way the IC receives feedback on the impact that operational action is having on the incident problem. These reports serve to provide a steady stream of ongoing critical info the IC uses to drive the tactical and strategic plan, and to evaluate the ongoing safety of the haz-zone workers. Effective and safe incident operations are the result of a series of actions that move through the standard tactical priorities (using the regular work cycle). The IC must connect what is being done currently with what must be done next to continue to move operations along. The two-way commo process between the IC and the workers is what keeps this ongoing process going. Operational safety becomes a critical part of the exchange of task- and tactical-level information. The organization must develop a system to routinely report on work status activities (i.e., is that work solving the problem?) as a regular part of each assignment. The IC must also

Command Safety

Communications

Safety Effect:

develop the capability to selectively ask for ("pull") specific information to maintain an awareness of how conditions are influencing the workers' welfare. The IC must also develop an effective sense of timing, to ask for the correct amount of such specific info at the appropriate time. The IC can then use that piece of info in the current stage of the decision-making process. Positioning and maintaining the IC located inside the CP "restrains" (and limits) the IC to that position (a good thing) and creates both the critical need for and the capability to deal with a steady stream of such progress, exception, and completion reports from all around the fireground. A major reason the IC can stay in the CP is that companies/sectors report (to the IC) what is going on in their position/function. If this doesn't happen, the IC must then go to those places (physically) to observe and evaluate. If the scene is very large or complicated, or if the situation is changing quickly, the IC must do some "aerobic orbiting" to try to stay informed. Orbiting ICs are not in any position to maintain overall control of the incident. The IC must gather and react to info on seven sides of a structural fire--the four sides, top, bottom and inside and what is going on in and around the layers of concealed spaces that come with each side. Lots of firefighters have died inside because that info was not quickly assembled, exchanged inside the incident organization, processed,

Safety Effect:

and evaluated, and in cases where the attack could not control the fire, the troops moved. The basis of maintaining this responsive capability revolves around info about critical factors being continually transmitted from right where those factors are occurring, and having that info being received and quickly processed (evaluate/consider/decide/order act by the command team). The IC is ultimately responsible for progress reports. If the IC doesn't receive what they want (specific information), they must ask. The entire team must continually exchange info on conditions in their area/function as an automatic process, so that a firefighter does not get beaten up/killed because someone had a critical piece of info and kept it a secret.

Conducting a standard critique at the end of the incident provides the entire team a chance to review how well the procedures/people performed. Commo is a critical part of that process. As this occurs over time, the entire team must become more skillful in exchanging high-impact tactical information in a way that is short and sweet = simple/effective... simple/effective = safe.

Command Safety

Communications

IC Checklist:

☐ Maintain your commo availability.

Safety Effect:

A major part of the overall safety plan involves maintaining an ongoing link between the hazard-zone workers and the outside (warm/ cold zone) world. The IC is the person who is responsible for managing this critical link. Radio partners operating inside the hazard zone also have their own special commo role. Companies/ sectors working in SCBAs must always be ready and able to hear a call from the IC. This is very challenging. Sectors and company officers must have the portable next to their ear--first call, immediate answer, inside and out (of the haz zone). When something happens that affects the safety of the workers, it is critical that the IC is available and positioned to quickly react--this reaction is generally performed over the radio. Being able to use the commo system to create a quick, effective response requires the IC to become adept and comfortable with using an audible (radio) method to assemble and process current, timely information and to create action that effectively matches the conditions that are seen and reported. The IC must make the personal transition from being a worker, who

Command Safety

Communications

Safety Effect:

creates action directly and physically right where the action is occurring, to being a boss who is physically separated from that action. This requires the IC to "quietly" give the orders that create effective and safe responses on the work level (particularly in the hazard zone). A major test of that transition (from worker to boss) occurs when there is a safety problem that requires quick action. This is the time when the incident needs strong command and control. If the IC hops out of the CP and grabs a tool to "save his guys," we have just added one more manual laborer and lost our only chance of providing the command-level (strategic) direction, control, and resources required to mobilize an effective rescue response. When a crew (or individual firefighter) gets stuck, they probably have lost their booming voice capability and may only have one small breath left to quietly gasp, "Mayday, Mayday, Mayday." Simply, if the IC is not in an effective "listening post" position (inside/quiet/good radios/concentration/focus/aides), the "gasp" may not be heard. If the IC can't hear it, the IC simply doesn't know about it, and the IC can't react to it. Establishing a standard system where the IC is always available to commo with is a major part of what actually having a strategic level of command really means. It sucks when old ladies with blue hair and sneakers call our commo center and tell them they heard "Mayday" on their Bearcat scanner, in situations where the IC is

Command Safety

Communications

Safety Effect:

on the roof showing a young firefighter how to start a chainsaw. In fact, this commo capability is a major part of having (or not having) an IMS. There is nothing genetically special about the IC. That person is the responder who arrived first and becomes IC #1 or the higher-ranking person who transferred command from whoever got there first. The commo system works because whoever is serving as a strategic-level IC for that incident is on a headset, instead of wearing a helmet (as the IC gains experience and age, the headset fits as well as the helmet).

Command Safety

Communications

IC Checklist:

☐ Utilize the standard order model to structure commo.

Safety Effect:

A major commo SOP involves using a regular method to send a message and then another regular method to verify that the message was received and understood. This very simple, standard commo process is called the order model. The beginning of the model involves the sender calling the receiver ("Command to E-1"). The next step is that the receiver acknowledges they are ready to receive ("go ahead Command"). Now the sender transmits the order ("do *XYZ* in the *LMN* position"). Now the receiver acknowledges they received and basically understood the order by transmitting back a brief restatement of the order ("E-1 copies, *XYZ* in the *LMN* position"). Using this two-way procedure (sender calls receiver/receiver acknowledges readiness/sender transmits order/receiver restates order) insures that the receiver is available and ready to listen before the sender gives the order, and then the brief restatement indicates that the receiver actually listened to and understood the order. Huge (sometimes huger than huge!)

Command Safety

Communications

Safety Effect:

commo problems occur when we think a message was received and, in fact, it was not. The sender (many times the IC) thinks the receiver got the order and assumes they are going to work on what the sender told them to do. Now the IC integrates the ordered (but actually unheard) task into the IAP and goes on to the next part of the plan. This is where the confusion begins, and many times this flub starts an unsafe "snowball" that begins rolling downhill. The company (with the flubbed-up order) is now many times out of position, doing something different from what the IC has written on the IC's TWS. This modest mistake can evolve into a major problem where that unit now has some safety need that the IC either is unaware of, or provides support to where the IC thinks that unit actually is (because that's where they were told to be by the IC) but not to where they actually are. The place to fix the problem is in the very beginning of the work-cycle process by using a standard commo procedure that effectively connects the sender with the receiver, and insures the receiver actually got the message that was sent. This becomes increasingly important when the action and fire are at their highest levels--this is when the IC can get rushed and has the "urge" to abandon the order model... don't! The IC must require standard commo procedures of everyone, and should model the use of those procedures when times are the toughest--even McDonald's requires a seventeen-year-old kid in a paper hat to use the order model (particularly when they are real busy)--"did you say large fries?"

Command Safety

Communications

IC Checklist:

☐ Keep commo simple: task/location/objective (use plain text).

Safety Effect:

Along with the order model, the team should attempt to always create commo that is very functional. Functional in an incident commo context means that the process must produce either an exchange of relevant information, or support in some way the actual performance of some physical action. This requires discipline and practice. There is a natural inclination to "chat" with each other over the radio--when this occurs, we begin to sound like a group of lonely truckers talking over their eighteen-wheeler CBs chatting about their last chicken-fried steak at Al's Truck Stop. How we manage (the details of) radio commo becomes a huge safety factor, particularly when the chips are down and we will revert back to our instinctive, everyday habits. Bosses should absolutely prohibit and react to anyone screwing around over the radio (really simple). The radio is as much a piece of critical safety equipment as a turnout coat, or an SCBA. The authors are not big fans of a lot of wasted effort on nonfunctional discipline (PFD senior staff dress uniform: Hawaiian shirt and jeans), BUT when there is a hazard zone present all the fun, informal, relaxed stuff stops and SOPs kick in. This is show time for the safety and survival of our troops, and this is where we had better be functionally disciplined and able to perform standard, safe operations. The IC must patrol the perimeter around the commo

Safety Effect:

process to be certain that valuable air time is devoted to coordinated work that is integrated into the IAP. The IC also sets the stage for supporting the IAP by keeping their commo simple--tell people what you want them to do, where to do it, and how it fits into the overall IAP. Workers operating in the hot zone should report back position, progress, and needs.

The senior author has been fascinated whenever he is in a remote location at home and overhears his wife tell the junior author, “Nick, please go into the other room and bring me the thing”... and then a moment or two later says, “Thanks.” My three children (all fire officers) and their mother (very beautiful) obviously have some special telepathic connection and capability. The IC very typically does not possess this relationship with the work units that must carry out the IAP--simply, they can’t read the IC’s mind--use the order model to tell them specifically where to go, what to do, what the objective is, and for whom they will be working. Emergency traffic or “Maydays” should be reserved for critical/urgent safety needs. The emergency traffic special tone should stop all routine radio traffic, until the IC indicates the problem is solved or stabilized. Plain English should be used throughout the commo process with emphasis on plain--we should leave the 204s, 301s, and 672s to Barney Fife. Everyone should use common words, language, and phrases. Save the poetry for the awards banquet.

Command Safety

Communications

IC Checklist:

☐ Utilize CP position and staff to improve commo.

Safety Effect:

A recurring theme in this essay involves placing the IC in a stationary, strategic position, inside a vehicle as quickly as possible. Among all the good reasons to do this, the most important is to protect the safety of the haz-zone workers--and a critical aspect of maintaining that safety is for the IC to always be in a physical position to be continuously available and able to commo with those troops. Keeping the IC locked up in the CP requires we create a decentralized system for supervising (and doing) the work all over the fireground with sectors and companies, and then using that same decentralized system to maintain a strong commo link with the IC. The IC must use the CP advantage to do this. The CP also provides a standard place (and hopefully space) for the IC to assemble a command staff who can assist with multiple-channel and TWS management (assignment/tracking/accountability), serving in standard IMS positions and doing assigned tasks for the team. Everyone in our business started at the bottom (the fun) and we all maintain a strong attachment to doing

Safety Effect:

task-level manual labor. It is a difficult transition for an officer who is a former foot soldier (all of us) to sit in an air-conditioned, well-lit, sound-proof SUV working as an IC watching hot/cold, dark, dangerous, dirty work (the fun) being done. We must continually engage in the reality therapy that shows that an incident that was not controlled by a fast attack, and effectively commanded by a mobile company officer IC, is now a continuing event that absolutely requires a strategic level of command; and it is impossible to achieve that strategic level until and unless we have an IC in a strategic position. Serious students of safety and command should not be distracted by the obsolete (but romantic) notion that the "commander in chief" should be in full battle regalia, complete with animal skin helmet (+ eagle) majestically standing in front of the fire--check the history--Napoleon eventually lost doing that (even though he had a good run) and got exiled to a small island without a good Italian restaurant. If you ask the troops (the sane ones) who are exposing their gazoo in the hot zone where they want the IC to be located, they will consistently and conclusively indicate they want the IC in a standard, strategic command position where it is possible to see, hear, maintain an awareness of their welfare, and to be continuously available to support and to help them move ("cover my back") if conditions worsen. In fact, the best place for the IC to be located to hear the troops is inside the CP where they pretty much can't be seen at

Command Safety

Communications

Safety Effect:

all. The old leadership model used the physical presence of the IC to visibly create an emotional symbol the troops could see and follow. The new IC in a CP model creates that same feeling and effect by packaging up the IC as a leader who is in a standard, strategic place (a command vehicle) which becomes the functional symbol of the human, physical, and procedural system that is in place that automatically and authentically protects the troops. Lots of firefighters have bit the dust in situations where the IC was surprised when they learned (fifteen minutes after it occurred) that something bad happened to their firefighters, who the IC probably never really had control of, because the IC never really had control of themselves (or anything else). The big deal, in all this, is that function should drive placement--not the opposite. This means (simply) that we want a strategic level of command (function) so we place that person in a standard strategic spot (placement)--big problems (safety/effectiveness) when we do the opposite. Serious command students should reflect on halfhearted CP routines and gadgets. Building a slide out CP unit in the rear of a van or suburban and having the IC stand at the back of the vehicle is still open-air, inhaling-exhaust, facing-the-opposite way (from the fire) command. Outside is outside. Having the IC operate on the other end of vehicle (the hood) is just a breezy. Rain, wind, dark, hot, cold, crowds, and noise all screw up the open-air IC. It's time for the American fire service to grow up,

Command Safety

Communications

Safety Effect:

bite the bullet, and get our collective command asses inside the rig (CP) where it's quiet, calm, light, and sequestered, and where we can do what the system pays us to do, act like strategic-level managers who focus on delivering wow service to Mrs. Smith, and then leaving with exactly the same number of firefighters (all okay) that we arrived with. The machine-shop manager ought to have worked "on the floor" to understand how the drill presses and milling machines actually work, but must play their role as a boss. This transition to being a boss requires they must be certain the electric bill gets paid (so the machines run), that the workers show up (so there is someone to run the machines), and that there is enough material (so there is something to work with). If that strategic-level boss can't say good-bye to their drill press, then they should stay on the task level and apply themselves to being a really good drill-press operator, so that someone who is prepared and willing to act like a strategic-level boss can actually perform that leadership role--sadly, the fire service still has a lot of plant managers (ICs) standing in front of their old drill presses because they sincerely believe that it is impossible to manage a machine shop unless you are up close and personal with your machine, with grease all over you, in the middle of a 105 decibel noise symphony.

Command Safety

Communications

IC Checklist:

- ☐ Center commo around tactical benchmarks--“all clear,” “under control,” “loss stopped.”

Safety Effect:

The regular rescue/fire control/property conservation tactical priorities make up the IAP. These basic priorities provide the foundation for a game plan that organizes and directs our efforts. This standard approach creates a safer operation because the players can predict and depend on what we are going to do first, second, and third. We also can attach a different risk management level to each priority (possible rescue--big risk/possible fire control [with property save]--little risk/anything already lost--no risk). Using the standard operational benchmarks of completion provides a standard way to indicate we have completed that priority and can go on to the next function. These priorities provide the framework for incident commo and become the performance targets for the incident operation. Fire companies must practice standard tactics (made up of standard skills) that match a standard order given by the IC in response to a standard condition, to (hopefully) result in a standard outcome (benchmark). The priorities create a system for the IC to assign specific units

Command Safety

Communications

Safety Effect:

to specific tasks in specific locations with standard (well known) objectives--achieving the benchmark for that priority. The standard benchmarks of completion-- "all clear," "under control," and "loss stopped" create a regular way to report that one priority is completed and we can go on to the next. The completion of the basic benchmark is a good place (in addition to standard-time intervals) for the IC to get a PAR on the hazard zoners. This greatly increases the safety of workers because it assists the IC in maintaining an awareness of the status of both worker safety and work progress. Tactical priorities standardize the plays and formations we use to perform haz-zone work, manage the risk, and the commo that is required to do this work safely.

Command Safety

Communications

This happens when Commo is done:

- ☐ IC uses commo to support, assist, and protect the team... commo is a high priority.
- ☐ Responders effectively get integrated into the incident organization and IAP.
- ☐ Everyone stays effectively connected to the standard commo cycle: condition, assignment, progress, exception, completion, and reassignment.
- ☐ IC gains info needed/IC knows and uses the standard info inventory for that situation to go after unknown critical factors.
- ☐ IC can control position/function of resources.
- ☐ Everyone can hear the IC--so there is continual command presence.
- ☐ IC is in a position to hear everyone and to effectively react.
- ☐ IC is always able to base IAP on the ongoing exchange of the most current, accurate information.
- ☐ The order model insures that standard, two-way commo is accurately exchanged between sender/receiver.
- ☐ Multiple channels kick in as incident organization expands.
- ☐ Changing conditions are anticipated/reacted to based on using ongoing commo to forecast and exchange info.
- ☐ Effective commo = launching pad for all eight standard command functions.
- ☐ Other commo support kicks in--multiple channels, cell phone, MDT, etc.
- ☐ Good IC commo helps others remain calm/effective.
- ☐ Accountability is maintained/safety is increased.

Command Safety

Communications

This happens when Commo is not done:

- ☐ Difficult to maintain an understanding on incident status.
- ☐ Difficult/impossible to hear/understand/reach IC--no command presence--no confidence that the IC is awake and mentally in attendance (sometimes physically).
- ☐ IC is ineffective... no effective info exchange.
- ☐ IC cannot effectively control the position/function of resources.
- ☐ Lots of unanswered radio calls--no response causes sender to "give up" on the receiver.
- ☐ Everyone becomes "disconnected"--lots of confusion and frustration.
- ☐ Steady stream of non-directed, uncontrolled, undisciplined, screwed up radio traffic.
- ☐ Difficult to maintain accurate responder accountability (either safety/ task).
- ☐ IC becomes disconnected from troops.
- ☐ Lots of free-lancing/hiding.
- ☐ Impossible to make quick (or really any effective) IAP changes.
- ☐ IC cannot effectively perform the other (seven) command functions.
- ☐ Info is lost/not received.

Command Safety

Communications

COACHING VERSION

"When you're the IC, capture control of the communications process from the very beginning by transmitting a brief, complete IRR. Set up command in a strategic CP as quickly as possible. Pay attention, listen critically, and quickly position yourself to always be available to respond to the first radio call. Always use all your CP commo advantages to continually direct and connect incident communications in a positive way to help folks trying to communicate from tough hazard-zone spots. Think of the CP as the "communications post." Effective communications becomes a major capability and mechanism (tool) of the IC. Effective, continuous commo capability becomes a major part of how the IC does all the regular functions of command. Remember, the IC pretty much is out of business if they lose control of the overall communications process. Always use the CP as your communications field office. Take your time, don't talk too loudly or too fast. You must develop the skill to listen critically "through" typical hazard-zone commo difficulties. Take a deep breath, sit back, relax, and stay in control. Always remember, the fireground is a lousy place to conduct any sensible commo... if the commo gets confused and out of balance, don't take it personally. Don't become impatient with commo difficulties. Quickly straighten out commo breakdowns and help the players effectively connect. Remember, you didn't start the fire--you're just tryin' to put it out. How you sound is how the troops evaluate the strategic level of command. Create a positive command image by being the incident radio superstar--don't ever let 'em see you sweat--sound cool--be cool."

Command Safety

Communications

Deployment

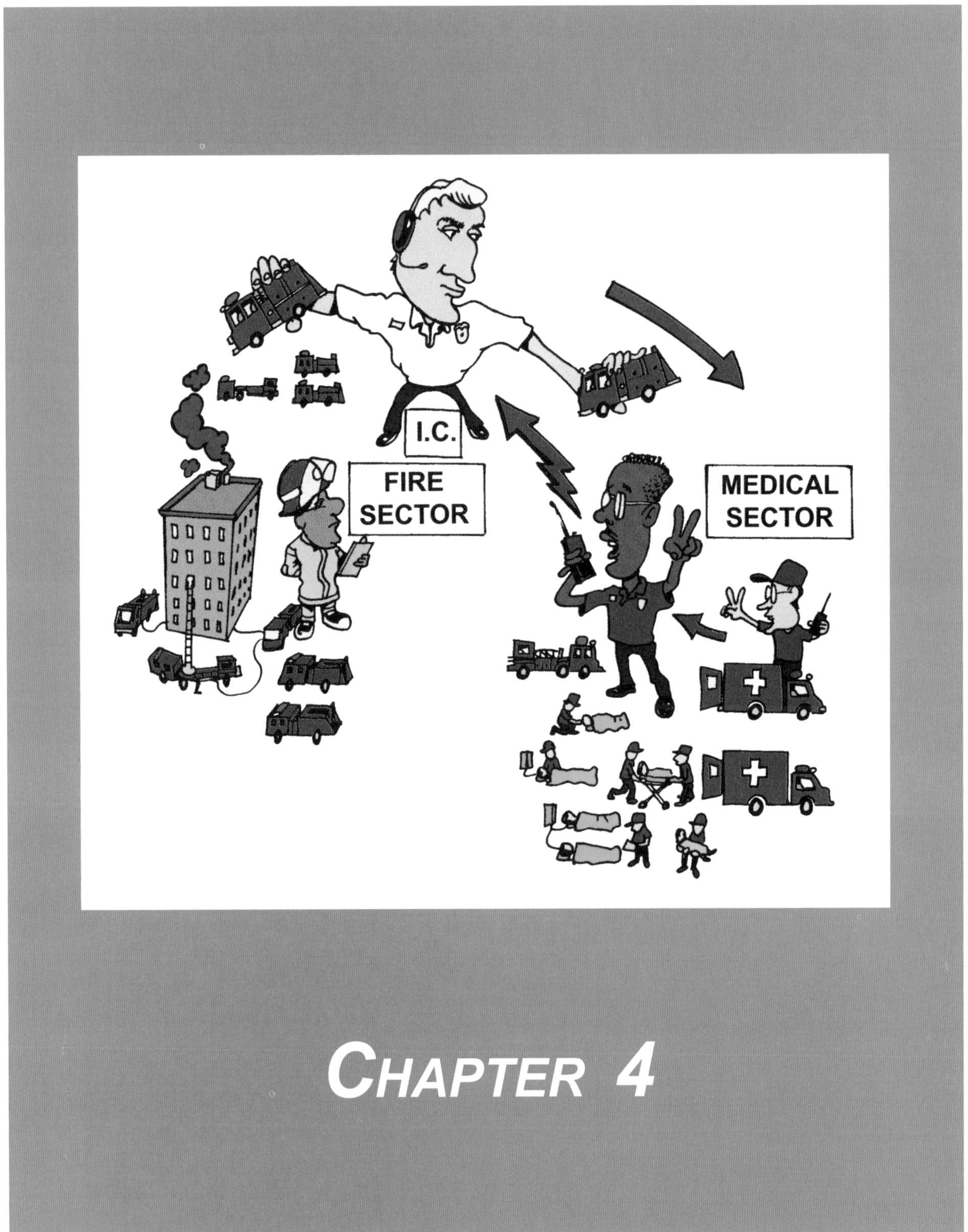

Command Safety

Deployment

Command Safety

Deployment

Function 4

Deployment

Major Goal

To provide and manage a steady, adequate, timely stream of appropriate resources.

IC Checklist:

- ☐ Call for resources based on the most rapid, accurate, current, and forecasted event profile you can develop and on standard tactical priorities.
- ☐ Maintain awareness of local response amount and capability:
    - personnel
    - apparatus, tools, and water
    - systems.
- ☐ Quickly access and use the local dispatch and status-keeping system.
- ☐ Monitor and manage within on-line response times.
- ☐ Use staging, assignment by the IC, and accountability SOPs to get firefighters into the standard work cycle.
- ☐ Maintain current, accurate, recorded, resource inventory and tracking on a TWS.
- ☐ Balance resources with task (don't overmatch).
- ☐ Always maintain an appropriate resource reserve.
- ☐ Use command SOPs to manage and escalate operations.

Command Safety

Deployment

IC Checklist:

☐ Call for resources based on the most rapid, accurate, current, and forecasted event profile you can develop and on standard tactical priorities.

Safety Effect:

The IC's deployment plan must be driven by the overall profile of the incident problem. The IC must quickly evaluate and estimate the severity and duration of the incident. These two factors (how big/how bad) make up the basic profile of how the event will evolve. The IC must use the event profile to evaluate and decide how much work needs to be done, how many workers (companies/sectors) will be required to do that work, and how long the event will last.

The IC uses the standard operational tactical priorities (rescue/fire control/ property conservation) as the basic performance targets (i.e., work plan) within the offensive/defensive strategy to safely and effectively manage incident scene operations. Using these regular priorities creates a standard order for what we do first, second, and third. Consistently applying this regular approach develops an understanding and confidence among the response team in what we will do and where we will go when we get on the scene, and (based on conditions) the acceptable risk we will take to do each priority. Using this standard game plan (over time) produces predictability and consistency. The

Command Safety

Deployment

Safety Effect:

IC uses the incident profile (for that particular situation) to call for the number and type of resources that will be needed to solve the incident problem and to complete the tactical priorities. Having the IC quickly make the call on the resource level, that will be required (using the basic event profile), provides an adequate number of workers to both get the job done safely, and prevents the first arrivers from becoming overmatched. The IC must call for a resource level that solves the basic incident problem, and a resource level that also provides for an adequate amount of tactical reserve (staged/RIC teams/resource pools/on-deck in rehab, etc.). Our work can be very dangerous. This rapid intervention tactical reserve must be automatically called for, assembled, and kept in place to deal with any urgent tactical needs that pop up and to be continuously available, ready, and in place to go into action, if firefighter safety becomes a concern. The type and amount of resources that will be deployed will become the deciding factor in what the incident organization looks like, the logistical response needed to support the attack, and the safety-support component needed to manage worker safety. A major deployment approach for the IC involves deciding on how much (and what kind of) resource will be needed, and then calling for that level of a response EARLY. In the old days, we had to call for additional resources sequentially (generally in greater alarms) because we had not yet developed staging procedures. If we did call for help simultaneously, those responders would quickly "cover up" the IC,

Safety Effect:

as they blasted into the incident scene (all arriving at about the same time), and automatically and independently went to work. What this (old time) "stretched out" resource ordering created was a delayed deployment reaction, that generally got behind (and stayed behind) the required resource curve, because the IC had to assign everyone who was already on the scene before they could call for more help (or risk getting covered up by the on-scene arrival of a mob of "unstaging" additional companies). The process consistently produced too little/too late (... the mother and father of disaster). The standard IMS work cycle creates the capability to proactively (rather than reactively) assemble the units we predict we will need early in the event to get them close to the scene (in standard staging positions/ staging area), available but not committed. The IC becomes the incoming resource "gate" and can then control the timing of the assignment of those resources within the incident organization and to the IAP, in a safe and orderly way. This creates a huge safety factor for the IC and the troops. It's a peachy deal to have staged resources located close to the incident scene who are available, quickly assignable, and ready to cover new positions, back up current ones, help those who are stuck, or relieve those who are plumb tuckered out.

Another modern deployment approach involves the automatic dispatch and response of the resources that make up the

Deployment

Safety Effect:

safety support system. The level of the safety response must match the size (profile) of the incident and the amount of units that were dispatched. Having these standard safety resources included as part of the regular dispatch process gets them on the scene quickly, and eliminates the need for the IC to remember (at the very worst time--the exciting beginning) to call for them. These resources can include RIC teams, safety officers, accountability officers, special operations specialists, senior safety managers, structural engineering support, special medical/rehab units and anyone and anything else that previous operations have shown to be needed to protect and support hazard-zone operations and operators.

Note:

The safety support deployment approach suggested in this section is a departure from the mentality we have used traditionally in how we dispatch resources. In the past, we have generally sent a single chief officer to manage each alarm... you get one command-level boss on the first alarm, one on the second alarm, and so on. We are beginning to understand that this outdated system simply does not provide the humans that are required to build and operate an effective and safe command and control response. As part of the original and ongoing alarm(s), these bosses (dispatched one per alarm) typically arrive on the scene too late to effectively capture control of a safe level of support. The dispatch system must beef up the

Command Safety

Deployment

Safety Effect:

command and safety system resources that are sent on the initial dispatch. The front end of our local operations are a highly dangerous period for our hazard-zone troops. This is a critical time where those workers must be protected by an adequate level of overall command and safety support. Old-time thinking was that we always had to be ready for the "next one!" Today we must provide an effective level of command and safety as early as possible in the incident, during the most active and dangerous period of exposure. We must then redistribute the units that are available to protect the rest of the jurisdiction in behind (and around) the event that is currently (and actually) underway. The operational welfare of our members becomes a compelling reason for us to develop response agreements that utilize regional resources in an area-wide, automatic-aid deployment system. There cannot be any higher priority than providing safety support to hazard-zone operations. Having responders sitting in recliners in fire stations watching the latest installment of the televised dysfunctional dramatics of those who are seriously relationship disadvantaged (Jerry! Jerry! Jerry!), when there is an urgent need to provide tactical support to a hazard zone that is two miles away, no longer is acceptable or explainable.

Command Safety

Deployment

IC Checklist:

☐ Maintain an awareness of local resource response, amount, and capability:

- personnel
- apparatus, tools, and water
- systems.

Safety Effect:

The IC's capability to pull off the IAP hinges on the ability of the deployment system to deliver the right amount and type of resources that are required to solve the incident problem, and to be able to deliver those resources within the problem-solving intervention time. The IC has got to be able to use the standard work cycle (responding/staged/assigned/rehabbed/reassigned/demobilized... sent home) to quickly get the right resources (people/equipment/systems) on the scene, effectively and safety in place, and to use operational and command SOPs to manage where they are, and what they are doing. It doesn't make much sense to bring all these players together, if the IC isn't able to incorporate them into the plan, and then effectively and safely manage them throughout the incident (... if

Deployment

Safety Effect:

you can't protect them, don't call them!). If the incident problem is a five on the ten-stage event scale, the IC will not take control of the incident problem until at least a level-five resource level and organization is in place. Developing an event profile is not an absolutely precise process, plus conditions can change quickly, so it's always smart for the IC to call for a little more resource than what is absolutely necessary. Having a level-six deployment response, at a level-five event, causes everyone to be safer because the IC has a little extra resource "cushion" that can be assigned wherever it is needed, based on changing conditions (be very careful of ICs who chase a level-five event all night with a level-four-point-five response). The IC must maintain an ongoing awareness of how to use the local deployment system to produce an adequate response, and how long it takes to safely put that response in place. Defining the size, severity, and speed of the incident problem more clearly defines how the risk management level will be applied and what needs to be done to manage worker safety. Serious safety problems occur when the IC does not effectively use deployment system resources to get ahead of, overpower, and control incident conditions. In these cases, overmatched firefighters can find themselves in dangerous and unsupported positions. If the command and operational system fails to quickly move those workers away from those conditions, they can be injured or killed. The IC must understand and manage

Command Safety

Deployment

Safety Effect:

the following basic deployment system components to create safe operations:

- Personnel

Being able to provide an adequate number of workers consistently is, and always will be, the most critical overall deployment factor. Every incident problem is solved by some form of manual labor and the effect of this work is what causes out-of-control conditions to be converted to under control. The only things in our business that are automated are the transmissions and the automatic external defibrillators (AEDs)... virtually everything else is hand operated by teams of real, live (foot soldier) firefighters who physically rescue trapped customers, manage fire streams, and manipulate the structure. The IC must be able to assemble a sufficient number of workers to create a concentration of effort (and effect) while the incident problem is solvable. Delivering an army of workers to the scene after that intervention point is just interesting, because the late arrivers really are more witnesses of what happened, than workers who actually prevent those things from happening. Too little and too late continue to be the mother and father of disaster. Worker safety is directly dependent on having enough troops arrive quickly enough so they can create an effective concentration of force within the basic intervention (problem solving) window of operational opportunity. The

Deployment

Safety Effect:

IC must mobilize enough support for those workers to provide the logistics, command team direction and support, and on-line operational back up necessary to keep the attack going. Many times, our firefighters will attempt to do more work than can be safely performed, based on their aggressiveness and dedication. The IC must continually evaluate the work that must be done and call for the number of workers required to do that work. A big-deal command function involves the IC forecasting how many workers and bosses will be needed eventually to stabilize the incident--and call for that resource level as early as possible. A lot of times we call for help incrementally (piece meal), as the event evolves. The problem with this approach is there is never an adequate amount of resource assembled and assigned to overpower the incident problem. The effect of this is that we are continually behind the power curve and our tactical reaction is never big enough to overpower the incident problem. Many times, when this occurs, our workers are overmatched and they get beaten up. The IC must always evaluate the number of workers actually on the scene and match their assigned tasks to their capability. In some cases, the IC must create a defensive operation when an inadequate number of workers are available to control what might otherwise be an offensive situation, if more workers were available, or if the workers could simply get there more quickly. Making this defensive call is a tough job for the IC because of the typical feelings firefighters have about directly engaging the fire. When such a "back off" decision is required, the IC must have the personal

Safety Effect:

stature and fortitude, along with smart, disciplined, and tough fire officers assigned throughout the incident organization to make it happen quickly, and then to control the troops within that decision. The IC must have the ability (and inclination) to engage the troops in "safety" commo (i.e., exchanging critical information that protects workers), as often as necessary. As difficult as this might be, it is far easier (and less painful) than attending funerals. The IC must demonstrate this preference whenever the incident safety level exceeds the capability of the safety system.

- Apparatus, tools, and water

Firefighters use their hardware in a variety of ways. They ride to the incident on BRTs and then put that apparatus to work doing such things as pumping water and providing access; and they always use the rig as a big mobile tool box, fire hose transport, and water discharger. Hand tools create the capability to physically move and manipulate the building and contents. Firefighters use hand and power tools to remove obstacles and barriers (doors, windows, walls, roofs, etc.) so they can put water directly on the fire. The IC must call for and coordinate the deployment of adequate physical assets so that firefighters can make continual forward progress and not get "stuck" in dangerous positions. A major on-scene deployment/safety objective is to create and maintain a variety of access options in and out of the hazard zone. When

Safety Effect:

we lose those movement options, the hazards in the hot zone become increasingly unforgiving, simply because the firefighter's ability to move away from those hazards becomes more restricted. Such a one way in/one way out access restriction must be reacted to and managed as a standard safety "red flag," and many times requires the IC to commit more resources and supervision to "open up" in a way that creates more access options. How the workers are equipped creates the capability for the firefighters to "defend themselves" and is a big deal in this access/safety process.

- Systems

A critical part of the safety program is the system that describes how we will operate in hazardous situations. We currently package most of that system(s) direction in SOPs. These local procedures give us the capability to plan and decide ahead of the incident what we will do and how we will do it--the procedures must also describe our limitations, where it is too dangerous for us to go, and what is too dangerous for us to do. These descriptions connect our risk management plan to actual operations. The SOPs become the foundation (starting point) of our basic management model:

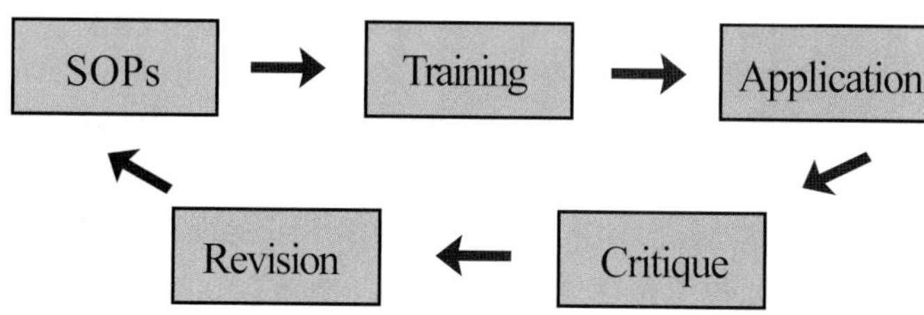

Safety Effect:

The application of this model becomes more critical to the management of safety than any other organizational activity. Based on the potentially severe consequences to our workers, safety SOPs are taught, applied, and managed as RULES, not guidelines, so the ongoing critique of home-town incident operations shows how we actually applied the safety SOPs (i.e., rules) at show time. As stated earlier, we encourage and promote inventiveness, and we should use a creative empowered, added-value based approach when we take care of Mrs. Smith's Siamese cat (customer service added value: people pets, pictures and pills)--on the flip side, skip the creativity when we are dealing with safety. Do what the SOP says all the time, every time--period. If it says stop, if the light is red, stop; if it says don't breath smoke, wear your mask; if it says stay with your crew, never be alone. If you are seeking an exciting career, join the circus and they can shoot you out of a cannon. If you are not smart enough to understand the difference between a rule and a procedure, have your head examined... you might be too dumb to be a firefighter (and if you are indeed too dumb, you are a hazard to yourself and everyone around you). Bosses must consistently use the critique to commend those who follow the safety procedures and to coach those who don't follow or have problems with them. Following standard safety routines takes a certain amount of extra time, and firefighters will many times cut corners, simply because they are in a hurry to do their jobs. Our experience reflects that the time invested in safety always pays off (and actually saves time) because we do it right the first time. We build positive safety habits

Command Safety

Deployment

Safety Effect:

by always paying attention to how we apply safety SOPs to our operations, and by absolutely following up (reacting) on our adherence to those safety SOPs. Lots of our members have been badly beaten up and killed because we didn't follow or enforce the standard safety routine we had agreed upon. It's dysfunctional and sad to have an accident we could have prevented, if we had just followed the procedure that is in our own SOP manual (lawyers have a love affair with these situations). Good bosses routinely deal with (and delegate) little stuff and sweat the big stuff, and there isn't any stuff bigger than developing, teaching, applying, and enforcing (commanding/coaching) standard safety practices. We keep repeating the term "habit" throughout this material. How we routinely, regularly, and automatically do things becomes a huge ongoing piece of the safety process (and outcome). Incident operations that are consistently performed within standard safety rules produce a set of functional habits. These habits cause us to do it right all the time, every time, everywhere, no mater if we are by ourselves or with a mob. These safety habits are what keep us out of trouble, and are the foundation of what gets us out of trouble.

Command Safety

Deployment

IC Checklist:

- ☐ Quickly access and use the local dispatch and status-keeping system.

Safety Effect:

The IC must be highly familiar with dispatch/ communications procedures and stay actively connected to the details of how that system works throughout operations. The commo center becomes an important incident management partner of the IC. They must use their initial customer contact, more powerful radios, commo center, office environment, direct access to on-line information on the details and status of the overall response system to assist and support the incident operation. The commo center knows what units are available, where they are, and directly controls the status keeping and dispatch system that can move and manage them. The IC must always use the IMS to get the right resources (closest to the incident/ appropriate type) in the right place, doing the right things. All these "rights" create safe operations. Effective commo is what dispatches the closest, most appropriate resources to create these safe operations. Our window of opportunity to solve the customer's problem (structural fire) is very narrow and is generally moving (i.e., going away) pretty quickly. The IC must be able to quickly deliver enough appropriate resources to interrupt what is

Safety Effect:

causing the incident problem, and then to create enough overpowering tactical force to cause the bad stuff to go away. The commo center fills the IC's resource request. The IC has to initiate and maintain the command system capability to keep track of all the resources assigned to the incident. If three engines and one ladder company (as an example) are assigned to the interior of a burning building, the IC had better be keeping track of where they are, what they are doing, what conditions they are working under, if they are okay, and who is going to help them out if they get into trouble. When we boil it down to its most simple and basic level, this is the major reason that the IC is sent to the scene. In most cases, if the IC takes care of the firefighters, the firefighters will take care of the customers. Having commo work in concert with the IC many times makes a huge difference in the overall command and control. A major deployment dynamic involves response discipline. Virtually every responder wants to attend any and every active incident that occurs (particularly a fire). This is a reflection of the terrific spirit and dedication of our members. Like most powerful feelings, our very strong response motivation has both a positive and negative side. It is a blessing for the customer and IC #1 to have a group of responders who bring their own motivation, personal drive, and dedication with them. This is good. Sometimes, these same feelings will cause them to monitor the dispatch frequency and to respond when they were not dispatched. This is bad. The foundation of incident operational control is

Deployment

Safety Effect:

that the IC must be aware of the general status of every response unit that is going to the incident, must log them in, assign them, track and account for them while they are in the hazard zone, and effectively rotate them to be certain they are adequately supplied and rehabbed. This simply requires the IC to know who is assigned (i.e., dispatched) to the incident, and then who has arrived on the incident scene. The IMS is absolutely outperformed (and sometimes overwhelmed) when companies and individuals self dispatch, arrive on the scene and just go into the hazard zone with no contact with the IC (a gigantic violation of safety SOPs). The bigger the event, the more likely that this self-dispatch process can happen. Massive confusion has occurred at recent major incidents when responders from inside and outside the local response agency have simply packed up and gone to the "big one" to "help" (sometimes across country--literally). These situations create huge clusters that are many times impossible to bring under control... sometimes it takes (literally) days to sort them out. The reason this occurs at the big ones is that those same responders self dispatch during regular times (at little ones), and no boss ever deals with them ("firm coaching" by the IC immediately after the incident). It is the IC's ongoing responsibility to call for an adequate deployment level and then to manage the level they have called. It creates huge safety problems when the IC thinks there are four engines and

Safety Effect:

two ladders on the scene (because that's what was actually dispatched) when, in fact, there are seven engines and three ladders, because the self dispatchers felt underutilized and just mounted up and "took in" the incident. The answer to this problem is really simple:

- ☐ If you are not dispatched, DON'T GO TO THE INCIDENT.

- ☐ As you approach the scene, you must automatically stop in a standard level-one or level-two staged position and report your arrival (according to SOPs) no matter if your being on the scene is "legal" (dispatched) or "illegal" (self assigned). You must then hold your staged position until you are assigned by the IC.

Command Safety

Deployment

IC Checklist:

☐ Monitor and manage within on-line response times.

Safety Effect:

The IC must be highly familiar with the regular, day-to-day location and type of local resources, and the availability of those resources. The IC must be able to quickly trigger the response of those resources by being able to effectively contact the commo center by using the correct deployment SOPs that articulate those resource needs. The IC is playing an application of force game with the incident problem. That game requires calling the right number of workers, and then having a general knowledge of how long it takes to get them to the scene. If the IC determines the incident problem to be a level five, the IC must then factor in how much time it will actually take the standard work cycle to put a level-five response into place, ready to go to work. By the time the IC assembles a level-five response, the problem may have escalated to a level six (or maybe a level that is even a lot bigger number). This requires the IC to effectively forecast where the problem will be when the calvary actually arrives (and gets set up and is able to go into operation), and then orders enough resources to handle the problem that will then be present.

Command Safety

Deployment

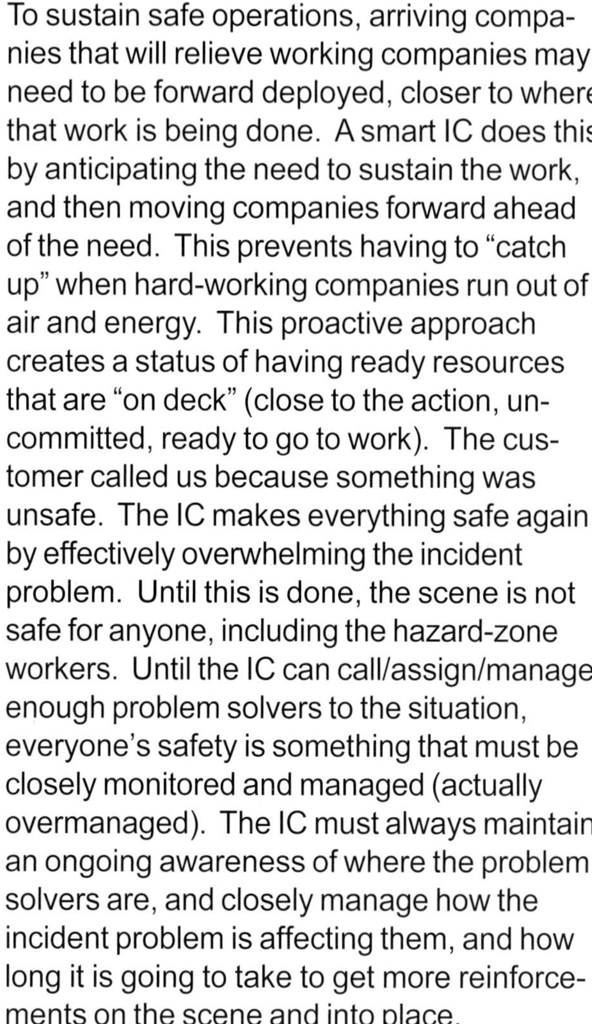

Safety Effect:

To sustain safe operations, arriving companies that will relieve working companies may need to be forward deployed, closer to where that work is being done. A smart IC does this by anticipating the need to sustain the work, and then moving companies forward ahead of the need. This prevents having to "catch up" when hard-working companies run out of air and energy. This proactive approach creates a status of having ready resources that are "on deck" (close to the action, uncommitted, ready to go to work). The customer called us because something was unsafe. The IC makes everything safe again by effectively overwhelming the incident problem. Until this is done, the scene is not safe for anyone, including the hazard-zone workers. Until the IC can call/assign/manage enough problem solvers to the situation, everyone's safety is something that must be closely monitored and managed (actually overmanaged). The IC must always maintain an ongoing awareness of where the problem solvers are, and closely manage how the incident problem is affecting them, and how long it is going to take to get more reinforcements on the scene and into place.

The IC does some fairly simple "bookkeeping" activities to manage response times. When companies are dispatched, the IC simply writes them down on a TWS. Responders are generally dispatched in their order of arrival, based on the general location from which they start (the station). In systems with automatic vehicle location (AVL), the units will be dispatched from their actual location, so the IC must pay attention to the

Safety Effect:

response order, because the responders may not be coming from their normal position of being in quarters. The IC must have an overall awareness of the regular location of the resources in relation to the location of the incident. This creates a general understanding and expectation of how long it will take for those resources (generally in fire stations) to arrive on the scene. Company officers should report when they will be delayed (beyond their regular response time) so the IC can work that delay into the deployment plan. As companies arrive, stage, and report, the IC must check them in on the TWS. Doing such response time awareness/ management provides the IC with a practical system to effectively connect resources available, with the work that must be done. This time-oriented approach becomes the basis of not getting workers into hazard-zone places that are overmatched because the calvary was on the way, but not on the scene yet.

Command Safety

Deployment

IC Checklist:

☐ Use staging, assignment by the IC, and accountability SOPs to get firefighters into the standard work cycle.

Safety Effect:

A critical period for both worker welfare and operational effectiveness (closely related) is when responders initially arrive on the scene. How we start operations becomes the launching pad (sometimes literally) for the rest of incident. Simply, how we go to work will determine how the safety process will or will not occur. We begin to set up unsafe, free-enterprise reactions if the command system doesn't quickly establish IC #1, who effectively "checks in" responders as they arrive. A lack of this initial command and control becomes the consistent front end of free-enterprise reactions of responders, who arrive on an incident scene where there is not an effective IC in place. Anytime this "free enterprise" occurs, it disrupts the standard command and safety system. Those who have not checked into "command central" are out of control (by definition). The IC must use a standard management system to convert the status of arriving units from their response

Safety Effect:

mode, to being logged into a standard tracking system, and then to being assigned to a place in the incident organization and operation, with a boss who uses a standard accountability system to look after them. The IMS must always maintain the capability to change the status (location/function) of resources throughout operations to match changes that occur in incident conditions and the ongoing incident stages that occur throughout the event. Our workers develop a strong "forward-moving operational/attack momentum" and it requires training and discipline to quickly transition (evaluate, redirect, alter, change, move, make a u-turn, etc.) that focus and energy. The ability (and agility) to effectively make those changes becomes a major part of the definition of being "under control." Unfortunately, the opposite is also true: if the IC cannot change the position/function of the hazard zoners, then the IC is essentially out of control. The command system uses a regular set of SOPs that establish the behaviors that our troops must initially use to safely go to work, and to continue to do that work, and then to make the changes that are required to safely get the job done. The application of the following SOPs become the foundation of a standard work cycle.

☐ Staging

Staging requires arriving responders to physically stop their apparatus before they complete the trip to the fire scene and go to work. This pause in their response gives the IC time to evaluate conditions, develop an

Command Safety

Deployment

Safety Effect:

IAP, and assign arrivers to their part of the plan. Using staging to create this standard entry "gate" results in a managed and safe beginning of the work cycle. In small/medium sized structural fire situations, most local (day-to-day) staging SOPs call for the first engine and truck company to go directly to the scene (generally side A) and begin standard company functions--all the other responders who arrive after the initial-attack team follow standard level-one staging procedures, which are pretty simple:

- Don't drive directly to the scene.
- Stop short--about a block away on a hydrant (if you're an engine).
- Transmit your staged direction (from the scene)-- "E-1, North."
- Wait for orders from the IC.
- Stay in your staged position until you receive those orders--unless you see urgent tactical needs... if you respond to those needs--tell the IC.
- Listen to and look at what is going on--pay attention.

We use another mode of staging (level two) for bigger deals that will require the response of lots of companies.

Command Safety

Deployment

Safety Effect:

Level-two staging is also simple:

- IC recognizes the need for lots of companies.
- IC declares level-two staging (commo center repeats this message).
- IC selects and announces staging area (parking area close to the scene with good in/out access).
- Now all responders quietly (no radio traffic) go to staging area (not to the scene).
- Companies park in standard staged position for quick exit; crew stays together on rig ready to go to work.
- First-arriving officer becomes staging sector officer, may be relieved later by other command helpers.
- Company officer "checks in" (face to face) with staging officer as they arrive.
- IC commos with staging officer to get companies into operating positions from staging.

Command Safety

Deployment

Safety Effect:

- It makes sense to use a separate channel (if available) between staging and the IC to keep the primary tactical channel clear for operational hazard-zone communication.

Level-two staging streamlines the arrival and assignment of a big response, and creates the capability to have a large amount of resources close to the scene, who are available to quickly assign within the incident organization, based on requests for resources from sector officers to the IC. Level-two staging has been a "quiet" IMS change that has created a huge command advantage--the IC can call for a big response and never has to really deal with those staged resources directly, until the IC is ready to assign them. All the level-two stagers stay off the radio. When they arrive at the staging area, they report in person (face to face) to the staging officer and don't in anyway make the IC nuts with a lot of radio traffic.

Both levels of staging create a sensible beginning for operations and replace the old-time process where arrivers would zip by the IC on their way to the fun zone... sometimes the IC caught up with and was able to control those free-lancers, and sometimes the IC never caught them. In those wild and woolly non-staging times, maintaining an awareness of the status of the resource (who/where/what) could be a big mystery for a long time. Staging is much safer... and we do a lot better job operating in the fun zone because the IC told us to go there. The IC can provide

Safety Effect:

the safety and support required to keep having fun, particularly if conditions get bad and the IC orders us to continue having fun Snickering up in rehab, while the fire building falls over and bounces about.

☐ Assignment by the IC

A major safety practice involves the IC making specific assignments to responders as they arrive on the scene. This integrates the workers into the incident organization and the IAP, from the very beginning of their time on the incident site. The standard assignment process is the command and control cornerstone of everything we do... the command system is not in place without it. Responders respond to work, not to wait around for orders (or we would call them waiters). So the longer they go without receiving instructions, the bigger the chance they grow weary of all the IMS stuff and will assign themselves. For this command assignment process to consistently occur, there must be someone to give the order (IC) and someone to receive and carry out that order (fire company/sector). How this simple (sender/receiver) process actually occurs becomes the most primitive way to evaluate if an IMS is, or is not, in place. The IC cannot manage (capture) and maintain an on-line awareness of the initial and ongoing status (position/function) of the troops, if they don't know where they are, what they are doing, and when they last spoke to them (brilliant analysis). The most effective way (i.e., basic foundation) for this control to occur

Deployment

Safety Effect:

is for the IC to assign specific companies to specific spots with specific objectives. To make such assignments, the IC must evaluate conditions, develop an IAP to deal with those conditions, and then translate that plan into orders the receiver can understand and carry out. This requires the IC to be conscious and thoughtful enough to produce and transmit orders that make sense to the workers, and to develop and extend those instructions quickly enough to make an impact on the incident problem. If we expect the workers to enter the hazard zone (which we do routinely), we should expect the IC to create and commo a doable/survivable work plan to those workers, and to use those initial assignments as the basis for maintaining control of the troops throughout operations.

Our service uses the ongoing application and refinement of SOPs to help the IC with the assignment process. As we learn what is effective by applying those SOPs in recurring situations, we are able to develop standard tactical routines (and recurring patterns) that effectively and safely match those conditions. Those experience-based routines are then translated into local SOPs. The SOPs describe where we go and what we do, based on our arrival order, in particular situations. This routine could be as simple as a standard (SOP) routine for a structural fire:

- First arriver becomes IC #1--begins the functions of command.
- First engine goes to side A.

Command Safety

Deployment

Safety Effect:

- Second engine "backs up" first engine.
- First ladder does standard ladder functions (forcible entry/vent/provide assess).
- Third engine takes side C.
- First BC transfers command--stays in vehicle--continues to perform the command functions.
- Everyone else stages level one.
- IC assigns balance of the responders.

These procedures become a huge help to the IC because they describe how we will do the beginning, routine (front end) part of our response. When we use these SOPs, particularly in the hazard zone, sectors and company officers should report to the IC that they are "in position" (according to SOPs) and are operating. The IC can quickly acknowledge and record the location, and note the function on the TWS. This SOP-driven approach provides a game plan that describes the standard action we will take in response to standard conditions. Doing this creates a lot "quieter" command process and frees the IC up to pay attention and help responders be safe and effective, by coordinating the activity that inevitably

Deployment

Safety Effect:

occurs "in between" the SOPs based on the particular needs of each situation.

Another major part in the assignment (and safety) process involves the method the IC uses to initiate operational action. What actually causes operational action occurs along a scale the IC must become familiar and skillful with. The scale ranges from the unit being assigned by deciding themselves what action they will take, to the IC giving a direct order. The severity of the hazard and the possible negative consequence (to our troops/ customers) that goes with that assignment will determine which option on the scale the IC will use to initiate action. Consistently making this connection between the possible safety/welfare consequences of a position or function, and the best method of assignment (i.e., safest), requires the response team to develop and apply an understanding and agreement on how "orders" will be given, received, and reacted to under actual incident conditions. This understanding and agreement must occur ahead of the incident, applied during the incident, and reviewed and refined after the incident, based on their actual application. Methods of direction ("orders") occur on the following scale:

basic individual values	empowerment	SOPS	IAP assignments	collegial interaction	direct orders

Deployment

Safety Effect:

Basic Values:

Basic values are the powerful and durable early teaching/ leaning/socialization gained from family experiences (both personal family and fire department family) about what is right, what is wrong, and how to treat others. Mom's fundamental teaching about basic citizenship, civility, and being nice = the personal ability and natural inclination to always "do the right thing." These personal values become the basis of how a person will act/behave when no one else is around. These are the "orders" we give ourselves.

Empowerment:

Delegation of authority allows the individual responder to perform within organizational SOPs, guidelines, and values without asking for permission. This pre-permission attempts to create decision making and action that is developed and applied by the department member closest to the incident problem and to the customer. It is fast and non-bureaucratic, and attempts to take advantage of the capability and dedication of each department

Safety Effect:

member. This method of direction requires a strong, supportive organizational context, and becomes a reflection of how members are "raised" and managed.

SOPs :

SOPs describe a regular organizational response that we apply to deal with a set of operational needs that typically occur at most incidents. This pre-event agreement creates the ability of a team of responders to perform these standard procedures automatically (without receiving orders). SOPs are applied to recurring tactical and operational conditions, and free up the IC and the command team to focus on making and managing hazard-zone assignments and operations. We use SOPs to decide upon, package up, and perform our operational safety routine. The ongoing application of those procedures creates the moves/formation/foundation of effective safety habits. SOPs must be written in a standard format and become the starting point for the SOP/training/application/critique/revision management model.

Command Safety

Deployment

Safety Effect:

IAP Assignments:

The IC initiates and maintains the IAP by making specific assignments to task-, tactical-, and strategic-level responders to the positions and functions that create the organization that is assigned to carry out the IAP. These orders also form the basic incident organization. They provide the direction to get us going and to keep us going because they get the workers (companies/sectors) initially in their assigned work spots. IAP assignments typically get responders into their basic positions where they and the IC will then normally use a collegial form of two-way communications to continue and complete the IAP, and to keep everyone connected. The IC uses the best, most current information to determine the overall offensive/defensive strategy and then a related IAP. Using this standard decision-making format (offensive/defensive) is simply the best initial shot the IC has to safely get the incident operation up and operating. Sometimes what looked like an offensive position (and assignment) from the front is, in fact, a very defensive one when the company/

Safety Effect:

sector gets to the rear. In this case, the company/sectors safely position themselves and call the IC back on the radio and describe conditions. Initial assignments must be used by the IC to geographically and functionally cover the entire incident. They are not absolute "marching orders" to operate as ordered, if that order is not safe and standard. This works really well when there are well-trained, rational grown ups on both ends of "the radio," who completely understand the IMS and safety plays.

Collegial Interaction:

This approach utilizes the skill, experience and actual physical positioning of the assigned unit, along with the CP physical position of the IC, for maximum team effectiveness. This method involves the active exchange of information and the mutual determination of the best action to take. The IC says such things as, "What have you got up there?" and "What do you think we should do?" This conversational method works well, takes advantage of having a capable, willing and highly cooperative team (eyes, ears, guts, brains) assigned all over the incident,

Safety Effect:

and is the method of information exchange, decision making, and intervention we use the most. Lots of times the IC starts out the interaction by transmitting a statement followed by a question mark (examples: "We need to begin topside operations, roof sector; how does that roof look? We need to get a line in through side C, E-1, can you make entry from the rear?").

Direct Orders:

Sometimes the IC wants an immediate response with no conversation. This need is based mostly on the need to quickly move the troops because of some urgent safety or welfare problem. This category of direction sounds like "get off the roof!" or, "abandon the building!" Lots of times the sender will request "emergency traffic," get the special tone/signal, and then transmit a direct order. Emergency traffic is designed to create an increased level of attention and urgency. Given that we normally use a conversational type of collegial interaction, when the IC gives a direct order (typically in a little bit different tone of voice), those on the receiving end should regard

Command Safety

Deployment

Safety Effect:

this change (style/tone) as a special signal and quickly respond. Based on this, smart ICs should reserve direct orders (statements always followed by an exclamation point) for urgent, safety-based needs.

Note:

Our natural operation momentum is very active so most of our communication is directed to “going forward” (i.e., attack). This approach typically involves the IC assigning companies to the basic positions and functions that get the operation going, and then engaging in ongoing two-way (collegial) commo to keep it going. Most absolute direct orders are then used to “go backward,” i.e., retreat. The vast majority of responders do not require big-time orders to attack--they do require them when they are in the attack process, and the IC decides their position(s) are not safe. An old IC said, “I never have to push ’em in, I’ve only got to pull ’em out.”

☐ Accountability

Taking the trip to the hazard zone requires that we have a system in place that always accounts for who is working in that zone, where they are working, and if that team is intact and okay. The accountability system is the entry “ticket” (pass) to the fun zone--no ticketless travel (free-lancing) allowed. The system also provides the accountability

Safety Effect:

verification that is used when we must quickly exit a hazard zone. Under difficult incident conditions, the IC must utilize the same system that was used to get the troops in, to quickly get them out. We should think of the accountability system as a transmission with a forward (in) and a reverse (out) setting. The accountability system is designed to show the names of the individuals on the team, along with their company designation. Career fire departments mostly use a "passport" system (velcro pad with plastic name tags) that always follows the company to their current assignment. Passports are routinely carried on the dashboard of the rig, right in front of the company officer. Volunteer fire departments mostly use personal tags that are attached to the firefighter. The tags have the firefighter's name or number and are collected together to track the workers. The basic mentality or discipline for the accountability system is that the company goes in (to the haz zone) together, comes out together, and must always stay together. The company officer must always be able to account for their members. The accountability system must operate on the strategic level by the IC who tracks companies on a TWS, on the tactical level by sector officers and accountability officers, who manage the passports outside the hazard zone, and on the task level by company officers who must always maintain vision, voice, or touch contact with their members inside the hazard zone. As the incident (fun zone) gets bigger, the IC must escalate the accountability system by setting up multiple accountability locations and

Deployment

Safety Effect:

accountability officers at those locations. The system uses the term PAR to indicate the team is together and okay, and at predetermined times requires the IC to get PARs from all haz-zone companies. Any "NO PAR" indication requires the IC to initiate the lost, trapped, missing firefighter procedure (big deal).

These are the IC responsibilities for a reported lost, trapped, or missing firefighter:

- ☐ Emergency traffic--as soon as the IC receives notification of a lost, trapped, or missing firefighter, they request emergency traffic to notify all of the response team that a firefighter(s) is in trouble.
- ☐ Change the plan (IAP) to a high priority rescue effort--the IC must adjust the current IAP to prioritize the rescue effort.
- ☐ Immediately request additional alarms--the IC must get the required resources (medical, heavy rescue, personnel, etc.) to the scene as quickly as possible to implement the rescue IAP. A level-two staging location should be established so the IC can assign these incoming resources according to the plan. Extra command-level officers will be required to upgrade sector officer positions, manage the tactical

Command Safety

Deployment

Safety Effect:

requirements of the rescue effort, and support the IC/ command team.

- ☐ Fireground accountability--the IC should concentrate first on establishing who is in trouble, the nature of the trouble, and their location. Once this is determined, the IC should get PARs from all of the other hazard-zone sectors/crews/ personnel.

- ☐ Assign a safety sector--the safety sector is responsible for identifying any safety hazards, monitoring activity, and evaluating the safety of the operation.

- ☐ Commit the RIC--command should send the RICs to the most appropriate location to begin the search and rescue effort. RIC crews should have the basic tools they will need to carry out their mission. This generally includes an attack line, TIC, and extra air for transfilling. The IC must assign however many RICs that will be eventually required to finish the search and rescue effort. In many cases, the initial RIC (IRIC) may only have enough air to locate the lost, trapped, or missing. If the IC hasn't forecasted ahead and sent

Safety Effect:

another RIC to that location, the rescue will break down, and the IRIC may find themselves in another Mayday situation. The IC should plan for a relay rescue from the onset of the operations (this requires multiple/staggered RICs).

- ☐ Coordinate the placement of preexisting units--in many Mayday situations, crews that were already assigned and working are the ones that initially find the lost, trapped, or missing. The IC must receive PARs from these units and determine their status (are they okay?). The IC must then determine if these units are in positions that can facilitate and speed up the rescue operation. Factors that will effect this decision include the conditions in their area, the amount of air they have left, and their proximity to the last known location of the lost, trapped, or missing.

- ☐ Continued control efforts on the incident problem--the IC must develop a plan that protects the lost, trapped, or missing from the incident conditions. For structure fires, this often times requires reinforcing fire attack positions and providing vertical ventilation. Depending on the

Safety Effect:

situation, it may also require the IC to write off savable property, in order to focus on the search and rescue effort.

- ☐ Control access into the hazard area--these are very emotional events. One of the first things the IC must do is establish who is in trouble and where they are, and then determine the status of everyone else. If the IC doesn't quickly gain control of who goes into the hazard zone, by establishing entry control in all of the sectors, firefighters will free-lance their way into the building to help their buddies. This is an excellent trait that firefighters possess, but it is impossible to manage an event where this takes place (sometimes it's very challenging to manage when we are not in trouble--try to manage it when firefighters are screaming "Mayday" into their portable radio). The IC manages this by placing chief (sector) and safety officers on all sides where entry can be made into the hazard zone. This accomplishes several important things; it puts the required organization in places needed to manage the incident. It gives the IC an experienced set of eyes for recon info, and it places command

Safety Effect:

and control in all of the key operating positions.

- ☐ Set up treatment and transportation sectors--the IC must have emergency medical help immediately upon the rescue of any firefighters from the hazard area.

- ☐ Support work--truck companies will be required to provide any needed support work. This includes forcing entry, opening all exterior doors, lighting the exits, and vertical ventilation.

- ☐ Technical/heavy rescue--if the Mayday was a result of structural collapse, the IC will need the expertise and equipment of technical/heavy rescue teams.

- ☐ Control the media--the IC wants to avoid situations where the families of the lost, trapped, or missing find out about their loved ones while the event is going on. This can be very difficult with news helicopters, continually patrolling while they monitor our tactical radio channels.

- ☐ Ensure that dispatch monitors all radio channels--if for some reason a lost, trapped, or missing firefighter is on a

Safety Effect:

channel other than the tactical channel, command must be directed to that channel.

- ☐ Manage the strategy--RIC operations are high risk. The IC must ensure that the rescuers do not become casualties. If incident conditions have deteriorated to the point of an imminent catastrophic event occurring (total flashover of the entire interior, structural collapse, etc.), the IC must withdraw rescue and control crews to safe locations. The IC cannot kill twelve (as a random number) more firefighters in a vain attempt to rescue ones already lost. RIC operations are necessarily done under the most difficult, stressful, and dangerous conditions that we face. Based on that reality, the IC must manage hazard-zone operations in a way that always prevents having to save the firefighters with RICs.

The effective window of opportunity for Mayday operations ranges from seconds to a few minutes. How long will it take the IC to put the proceeding operation in place? Do you still think rapid intervention is rapid?

Command Safety

Deployment

IC Checklist:

☐ Maintain current, accurate resource inventory and tracking.

Safety Effect:

The IC's job is to always keep track of and control the position and function of all the incident players. This capability becomes a major safety factor and must be done within a standard, practiced system. The major tool the IC uses to do this is a TWS (and a pencil), along with a radio. We described TWS use in section two. The TWS has a regular place for the IC to list the responder's designation, when and where they have staged, where they are assigned, and a basic diagram of the incident site. As companies/sectors are assigned, the IC must note their location on the incident-site drawing. At the bottom of the sheet there is a basic set of organizational boxes that is used to show the geographic and functional sectors that have been established, and the IC writes in the company designation (in that sector box), as they are assigned to that sector. The TWS system is low tech, fast, and simple. It is the major way the IC maintains an ongoing inventory and tracking awareness of local resources (particularly, those in the hazard zone) of haz-zone worker status. Over time, as TWSs are used,

Safety Effect:

they become the standard place/process where critical incident info is recorded and formatted, so the sheet becomes a major tool the IC uses to record, manage, and exchange information. The TWS standard work cycle flow chart is as follows:

1. dispatched to the incident
2. responding
3. staged
4. assigned:
   - assigned by the IC (or SOP)
   - en route to their assigned position
   - working on their assignment
5. rotated through rehab
6. reassigned
7. demobilized (debriefed, if necessary)--sent home.

The consistent use of the TWS creates a positive habit and this habit pays off on the 500th (or 1,000th) call where the IC must move the troops quickly and conclusively, and a current, accurate work sheet shows who's where, doing what.

The work cycle creates a description of regular status categories that every responder moves through as they go from responding, to staged, to being integrated (i.e., assigned by the IC) into the incident organization and then working on the IAP. These standard (described in local SOPs) work-cycle categories become the inventory

Safety Effect:

and tracking foundation for everyone operating at the event, and become a huge safety factor. Notice that there is no category for self assignment (free-lancing)--you gotta' be in one of the regular places as you arrive, go to work, and continue to work. If you aren't in one of those standard categories, you are a "fugitive" from the regular safety system... when this occurs, the IC must act as the incident sheriff who rounds up the "fugitives" and places them under "IMS arrest" (in the regular incident organization).

Command Safety

Deployment

IC Checklist:

- ☐ Balance resources with task (don't over match).

Safety Effect:

The most consistently critical deployment element from both a customer-service and worker-safety standpoint is the number of firefighters we can deliver to the incident within a problem-solving time frame. Virtually everything we do at a structural fire requires quick, active (many times violent) manual labor performed by hand, by a completely non-automated service (you had better believe the International City Manager's Association [ICMA] would have automated us a long time ago, if they could). A major performance/safety problem occurs when too few firefighters show up at a fire that is actively underway, and effective fire control requires multiple tasks to be quickly performed (pretty typical incident situation). Based on their training, socialization, and feelings, firefighters will naturally attack just about any fire situation, because in the firefighter factory they build us to do just that--ATTACK!--we don't ponder, reflect, philosophize, appoint a committee or get in touch with our feelings--we attack. This absolutely admirable attack-oriented inclination can produce a highly-unsafe situation where six to eight firefighters will aggressively overmatch

Command Safety

Deployment

Safety Effect:

themselves trying to do a set of tasks that require fourteen to fifteen. Too few attackers create the following:

- Firefighters go into the hazard zone to directly engage the incident problem(s).
- Firefighters are now in dangerous, under-resourced operating positions.
- Supervisors stop supervising and must physically become primary attackers.
- Firefighters remain in hazard-zone positions longer than they should, and operate with dangerously low levels of SCBA air; they quickly lose their "round trip" ticket.
- Such overmatched attacks (and attackers) cannot create an adequate force to over power the incident problem(s).

- Now these firefighters are directly exposed to (i.e., close to) big, uncontrolled thermal/entrapment/ getting lost/collapse problems that are worsening.
- There are inadequate (or late) resources available to increase the size of the attack, back up the insiders, or provide for their rescue, if necessary.

Command Safety

Deployment

Safety Effect:

A major IC command function involves balancing the work that must be done with the response time and number of workers. Our service must establish a national deployment standard* for urbanized, developed areas that calls for:

- 4-5 person engine staffing
- 4-6 person truck staffing
- 1 response chief + chief's aide
- 5-minute initial-unit response time
- 9-minute 2 engines, 1 ladder, 1 response chief response time.

Response Time = time from dispatch to at the curb on scene.

Incident commanders must continually evaluate the on-scene work that must be done in relation to the number of available firefighters. In cases where an inadequate number of firefighters are in attendance, the IC must establish safe operating positions that in many cases involve writing off unburned property. This is a difficult process and requires tough decisions by the IC, and a disciplined response by the firefighters. There is no point in living in a short-staffed world and operating like we have standard staffing levels (listed above). In the real world, something must give, and it's lots better to make it the property and not the workers... for the IC, it's real simple--there is no building on earth that is worth the life of a firefighter (period).

* see NFPA 1710

Command Safety

Deployment

IC Checklist:

- ☐ Always maintain an appropriate resource reserve.

Safety Effect:

A current part of command system management involves the idea that the IC should maintain a tactical reserve of resources that can be quickly "plugged in" to critical places/ functions as the incident evolves, and safety and operational needs change. In the good old days, we had a very "full-employment" mentality. Traditional bosses were very concerned with anything/anybody that was not assigned and operating. If they encountered any nonworkers, they would quickly suggest(!) that the "slackers" find some manual labor to do or they would get a size eleven and a half BC boot in the butt. While this did create full employment and built character in the workers, it didn't work real well if something changed that needed a quick response. Many times when this happened, the IC had to strike another alarm and then wait (and wait, and wait, and wait) for those resources to arrive to do whatever needed to be done--which was many times too late (and put us in the middle of the dreaded catch-up game).

Now we use IRICs, RICs, responders in staging, and crews "on deck" (ready for assignment) in resource and rehab sectors,

Safety Effect:

as standard organizational elements that create and manage a tactical reserve. This way of establishing and maintaining extra, uncommitted resources that are in place, intact (crews together), and quickly available should occur automatically. The IC should list their location on their TWS. Many times in the beginning of a fast-moving, expanding incident, an IC will be assigning resources as quickly as they arrive. In these cases, the RIC team will probably be the only standby resource available. When this occurs, the IC should call for additional responders, if the forecast indicates the incident will continue to expand. There is no need for the IC to call for excessive resources beyond a sensible standby level. Having a little extra is much better than always being behind the power curve, and having to somehow desperately call for more responders that end up arriving after the problem can be effectively and safely solved.

Command Safety

Deployment

IC Checklist:

☐ Use command SOPs to manage and escalate operations.

Safety Effect:

The amount of labor that is required and the safety considerations that support that operational profile drive the resource level needed to control the incident problem. The IC uses the standard command system (i.e., command functions) and SOPs to escalate and manage the incident operation. If a level-five attack is needed to solve the incident problem, the IC needs to escalate to at least a level-five operational response and command organization (or ideally a little ahead of the incident profile). The IC must always maintain the capability to manage the position and function of all assigned resources, and must build a command organization that allows that control to occur.

A major challenge for the IC is to make the command organization (size/shape) match the work that must be done at the scene to solve the incident problem. Performing the regular functions of command is how the IC creates this "match." Everyone's safety at the scene depends on the IC being able to stay ahead of the incident problem--in offensive situations, the IC orders workers inside the hazard area to directly engage the fire. Offensive operations typically occur very quickly, so the speed of the attack and the required

Deployment

Safety Effect:

support becomes a major safety challenge. Defensive conditions require the IC to keep the troops outside and away from the hazard area (i.e., collapse zone), so they can limit the loss to the fire area (surround and drown), and save the rest of the neighborhood. Defensive operations can extend over a longer operational period (compared to offensive times), so the IC must establish an incident scene control approach that continually maintains safe positions outside the collapse zone. When the IC must make the operational response larger, additional resources must be called. This becomes a time where being able to escalate the command capability required to effectively manage the bigger response is particularly critical to worker safety. It's pretty easy for the IC to mash down the mic button and say, "Give me two more alarms." In those situations, if the IC doesn't escalate the incident organization to expand the standard command functions, we create a big, essentially uncommanded, unsafe mess with lots of potential victims that we hauled to the scene on BRTs. It makes sense for the IC to order command resources one alarm ahead to get command helpers on the scene and integrated into the IAP and incident organization, before the resources they will supervise show up.

Sending an adequate number of command team and staff must be built into the regular

Deployment

Safety Effect:

dispatch process, and it must be done automatically. Multiple chief officers must be dispatched on original assignments and must be a regular part of what is dispatched on each alarm level. In the old days, we used to send one chief on each alarm, where six to eight engine, ladder, squad, medic and special companies were going. If a second alarm was called, we got one more chief with the same response. Then we had two command officers trying to manage more than a dozen companies. Such overloaded (pre-IMS) officers were quickly swamped, and typically were never able to establish any effective level of control. Every local system must now superimpose the staff requirements to adequately establish and maintain their IMS, and then automatically include that number of supervisory (chief officer) positions in the original dispatch. Lots of current dispatch systems suffer "sticker shock" when they now evaluate and decide on how many command officers are actually required to build an effective command effort. This period of adjustment requires leadership and patience. In many places, response agreements among neighboring departments may be required to provide enough command officers to create an effective and timely level of support.

Command Safety

Deployment

This happens when Deployment is done:

- ☐ Commo center captures necessary info from the calling customer to determine effective/adequate dispatch profile.
- ☐ IC #1 develops a quick incident profile (size/duration... how big--how long) to determine and then call for deployment needs to meet tactical priorities; adequate resources are provided--IAP overpowers the incident problem.
- ☐ IC continually manages and controls resources within a standard deployment cycle: dispatched/responding/ staged/assigned (working)/rehabbed/returned.
- ☐ We use resources effectively.
- ☐ IC continually knows the location/function of firefighters.
- ☐ Workers are effectively protected by adequate command/support resources--not overmatched/not overkill.
- ☐ We control incident condition and interrupt/intervene in the incident problem(s).
- ☐ IC maintains staging/RIC tactical reserve.
- ☐ Customer stabilization gets completed effectively.

Command Safety

Deployment

This happens when Deployment is not done:

- ☐ Screwed up dispatch--too many, too little, wrong type of resources.
- ☐ Calling for too little/too late keeps us behind the incident/operational/safety "power curve."
- ☐ We continually struggle to catch up.
- ☐ We don't use staging/assignment/accountability.
- ☐ IC doesn't know who is where (and why).
- ☐ Incident conditions control us, overpower us, and eventually win.
- ☐ Everyone gets frustrated--firefighters want to shoot IC.
- ☐ Safety is compromised--overmatched workers work too long/hard, and get beaten up.
- ☐ We cannot effectively protect Mrs. Smith and her stuff.
- ☐ Nothing is standard/everyday is a new day (deployment amnesia).
- ☐ Life sucks.

Command Safety

Deployment

COACHING VERSION

"Figure out what you will need, based on what is going on, and develop your best shot at what will happen, and then order the resources you think you will need. Don't screw around--be pessimistic, order big, order early (send them home if you don't need them). As the IC, don't write a tactical check that the troops don't have the resources to cash in the hazard zone. Virtually every fireground operation requires real-live firefighters to execute. Most of the time the limiting factor in effective resource management and incident problem intervention is how many firefighters the system can deliver in the front end of the event. An army of firefighters who arrive after the problem-solving window of opportunity closes is just interesting. Use the regular work cycle SOPs, staging, assignment by the IC and accountability to get firefighters where they need to be to get the job done. Use those same regular system elements to keep track of them while they are in the hazard zone. You pick the time and the place where you're going to fight, and take enough with you to get the job done."

Command Safety

Deployment

Command Safety

Identify Strategy/Develop Incident Action Plan

Chapter 5

Command Safety

Identify Strategy/Develop Incident Action Plan

Function 5

Identify Strategy/Develop IAP

Major Goal

To use a regular, systematic method to make strategy decisions and to develop and initiate an IAP.

IC Checklist:

- ☐ Decide on the overall offensive/defensive strategy.
- ☐ Develop the overall strategy using critical factors.
- ☐ Apply the standard risk management plan.
- ☐ Declare the strategy as part of the IRR.
- ☐ Manage and control operations within the basic strategy.
- ☐ Implement an IAP to match the overall strategy.
- ☐ Use critical factors to develop the IAP.
- ☐ Include strategy, location, function, and objective in IAP.
- ☐ Use tactical priority benchmarks as action-planning road map.
- ☐ Reannounce ongoing strategy confirmation as part of ET reports.
- ☐ Do not combine offensive/defensive operations in the same fire area (compartment).
- ☐ Use the incident organization and communications to connect and act out strategy/plan.

Command Safety

Identify Strategy/Develop Incident Action Plan

IC Checklist:

☐ Decide on the overall offensive/defensive strategy.

Safety Effect:

The word "strategic" is used so much throughout this essay because it is so important to safe and effective operations, particularly during structural fire fighting. The entire team uses the initial and ongoing strategic decision, as the basis (and beginning) for where and how the fire fight will occur. All tactical decisions and operational action emerges from the strategic decision. Offensive/defensive tells us operationally where we go, what we do, and where we don't go, and what we don't do. The "go"/"no go" approach to hazard-zone entry and operation is a very simple and practical way that both describes the general conditions, and the specific reaction we take to those conditions... basically, we can go any place where our safety system can adequately protect us and provide a round trip into and out of the hazard zone--we must not go anywhere where the hazards are bigger than the safety system (simple to say; hard to do). The strategy identifies how companies/sectors are going to work together to achieve the tactical

Command Safety

Identify Strategy/Develop Incident Action Plan

Safety Effect:

benchmarks. Making the overall strategic decision requires the IC to evaluate the critical incident factors that are present and forecasted, and then translate that ongoing evaluation into an IAP that determines and describes the basic details of our critical operating positions/functions. The operational and safety plan must be developed around SOPs and regular IMS procedures, so it can be quickly understood and carried out by the troops.

We hear a lot about "split-second decisions" in our business. While this sounds exciting and dramatic, the "split-second" mentality is pretty scary when we consider the possible (fatal) consequences of the tactical decisions, that are many times attributed to one of the "split-second blinding flashes" of instant decisions. We had better base how we decide what to do (particularly in the hazard zone) on a huge pre-incident preparation approach. SOPs and IMS together create a rational front end (i.e., foundation) to creating and then managing both the overall strategy and the IAP. Having the organization decide way ahead of time how operations will be conducted in standard situations (which are most of what we have) becomes the very practical basis for quick incident scene decision making and operational action. This front-end "preloading" on standard operational protocols, the application of lessons learned and reinforced in past operational experience, and personal, smart intuition support the IC when they must indeed make what looks like to the outside world a quick ("split-second") decision. This

Command Safety

IDENTIFY STRATEGY/DEVELOP INCIDENT ACTION PLAN

Safety Effect:

decision-making preloading is not meant to in anyway discount or disrespect the fast, critical decisions the IC (and everyone else on the team) must make during incident operations. The proactive (discuss, decide, practice, refine ahead of the incident) approach actually supports this rapid decision making reality. Any part of our operational process that can be decided ahead of the incident, loaded into our IMS system, practiced ahead of time, and then automatically applied on-line, gives the IC more time/attention-span capability to make critical, rather than routine decisions. A major part of this approach involves the development of a highly refined safety system made up of practiced, enforced, reinforced SOPs (that are managed as rules, not guidelines). This creates a huge support capability to every command/operational level because the whole team can concentrate on delivering standard service to the customer, and not having to save each other.

The IC must automatically apply the strategic-level safety routine as an integral part of doing the standard command functions--that is the very practical reason that the safety material presented in this essay is structured exactly around the command functions outline. Integrating the two (safety/command) eliminates the IC from making safety an after thought that gets remembered ten minutes after the incident is over (... oops!).

Command Safety

Identify Strategy/Develop Incident Action Plan

Safety Effect:

The development of SOPs, based on operational experience, becomes a major "organizational learning" method. We are cursed to making the same painful, old mistakes over and over again, if we don't consistently (and automatically) load our individual and collective lessons into the protocols that direct our incident operations. It's lots better to actively use the critique and revision approach for keeping SOPs current. Simply, if something we have done has (or can) hurt us in any way, we should change the standard (and sometimes special) plays that create that unsafe action, to avoid that pain in the future... the most important fire we will ever fight is the next one.

The IC must always realize the firefighter's very personal anatomy and physiology is being bombarded by incident conditions, while they are operating in the hazard zone (that's why we call it a hazard zone). The IC must develop a game plan (more applied preloading) that understands and balances the capabilities/limitations of the standard safety system, with a continual evaluation of the incident hazards.

The fire knows the firefighters are trying to make it go away. The basic demon-like behavior of the fire makes it want to continue to just naturally screw up the people and the stuff it can get close to, and then get on, so those extinguishment efforts make the fire really grumpy. The IC must realize that the fire (very grumpy) is always trying to injure/assassinate the attackers. The offensive/defensive decision

Safety Effect:

becomes a major, big-deal factor in extending the life span of the firefighters, who must go into the hazard zone to do their job.

The IC should not take it personally or think it a failure to go defensive. It is a major IC function to match the strategy to actual incident conditions. This is why we keep talking about standard action, connected to standard conditions, to produce standard outcomes. The IC (and the troops) didn't make the world--they are just trying to deal with the (many times combustible) conditions they encounter. When the safety score goes negative (fire/hazard is bigger than safety system), the IC must react defensively, and it really doesn't matter what (or who) caused the safety score to be less than the hazard score. When this happens, the IC must get (or keep) the troops out of the hazard zone. Strategic positioning is not negotiable (particularly defensive decisions/declarations). Big bosses (fire chiefs, assistant chiefs, shift commanders) must not only clearly state that a strong defensive reaction must be taken when the safety score goes upside down, but must commend and support the IC that actively acts out that defensive approach to protect the troops. How bosses react on the current fire will affect how the troops will behave on the next one.

Command Safety

Identify Strategy/Develop Incident Action Plan

IC Checklist:

☐ Develop the overall strategy using critical factors and the standard risk management plan.

Safety Effect:

The IC identifies the strategic mode as offensive or defensive through the analysis of an array of standard critical incident factors and their related characteristics. The major factors and questions to consider in determining the correct mode include (not in any order):

- ☐ Size--What is the size of the building that is on fire, the exposures, and the general fire area?

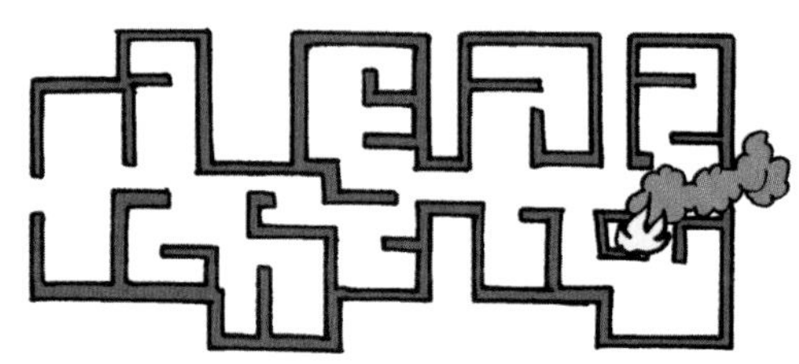

- ☐ Arrangement--What is the layout of the building/area, what is the profile of access, barriers, obstacles, congestion; what are the entry and exit characteristics?

- ☐ Fire extent and location--How much and what part of the building is involved? What is the fire profile from the burned (and burning) portion to the unburned portion. This factor will determine where we can and cannot strategically operate.

- ☐ Savable people--Is there anyone alive to save (rescue)? We will not risk our lives for those who are already lost. We must understand the difference between a rescue and a body recovery.

Command Safety

Identify Strategy/Develop Incident Action Plan

Safety Effect:

- ☐ Savable property--Is there any property left to save? We will not risk our lives for property (stuff) that is already lost.

- ☐ Entry and tenability--Will conditions allow firefighters to get in the building and stay in? Get in/stay in/get the job done/get back out = offensive; can't get in/stay in/get back out = defensive.

- ☐ Exit--Can the IC provide protection to the exit way (both exit and pathway to/from that exit) to maintain an ongoing way to get out of the hazard area? Two ways (or more) out are best. A continuous exit capability must be an integral part of the initial and ongoing interior offensive decision.

- ☐ Ventilation profile--Can roof operations be conducted? If we can't go/stay on the roof (long enough to complete roof ops), we should not go/stay inside.

- ☐ Special hazards--Hazmat, confined space, high/low angle, swift water, meteor showers? We must develop strategy/IAP to match the special hazard.

- ☐ Violence--Social disorder, shooting/bullets/weapons, violent people and situations are present--do we need law enforcement to go in ahead of us to stabilize, before we can safely operate? We (fire service) are basically a highly vulnerable, non-bullet proof "friendly force."

Command Safety

Identify Strategy/Develop Incident Action Plan

Safety Effect:

- ☐ Resources--Are sufficient resources available for the attack? We require an adequate number of manual laborers along with their trucks/tools/water to do incident work, either offensive or defensive. The IC must match the size of the work to the size of the work force.

Note:

The resource level that is actually on the scene and assignable is always a big deal in deciding on the overall strategy. The IC must be aware of and manage both the amount and the arrival times of those responders. Many times the IC is attempting to get ahead of a rapidly-moving window of offensive opportunity. Being able to control the incident problem inside that window requires quickly doing the work that must be done in that offensive-time frame. If this does not occur, the incident moves on (actually burns on) into the next "window," where the problem is rapidly expanding and moving toward (or actually inside) the defensive end of the scale. A big (huge) safety problem exists if the IC's IAP (and strategy) get behind, because the response was too little/too late, based on the current window. The response level that is assigned to the incident is a realistic reflection of the very local deployment "picture" at that point in time, and is in no way meant to be a criticism of the IC. All the IC can do is call for the resources that are available when the incident actually occurs. When too little/too late happens, it becomes the essence of being behind the "power

Command Safety

IDENTIFY STRATEGY/DEVELOP INCIDENT ACTION PLAN

Safety Effect:

curve"... being behind this very dynamic(!) process will always produce an increased level of risk to hazard-zone workers. What this means in street terms is the IC must order big/order early, hit the problem(s) as hard as possible, evaluate the effect of the effort, and then quickly react--if the problem is gettin' better, keep going--if the problem is winning--get out'a Dodge, set up outside, and live another day. Sometimes the IC can take control of the power curve (profile of incident conditions), and sometimes the IC can't. When the situation is a loser (for whatever reason), the IC absolutely cannot bet the firefighters' lives on an unforgiving situation, that has a life of its own. Making this "do I have enough resource" decision is a major reason that the IC gets up in the morning, puts on their navy blue costume, goes into work, and hops on big (or little) red. Having to make this call doesn't happen very often (thankfully), but when it does, the IC had better be packaged up and prepared to do so, or a couple of days later, we are all going to be wearing our badges and ties, listening to some sad, bagpipe music.

Command Safety

Identify Strategy/Develop Incident Action Plan

IC Checklist:

☐ Apply the standard risk management plan.

Safety Effect:

The following is a basic, risk management plan that effectively integrates into our regular IMS. The plan expresses a standard, three-level risk approach the IC must apply to our hazard-zone workers, who may be called upon to do dangerous, potentially fatal, incident work. The IC must use the standard, risk management plan as a major part of making and maintaining the overall strategic decision.

1. We will risk a lot to protect a savable life.

   This is a situation where we evaluate that there is a viable customer(s) threatened and trapped by the incident problem, and where a physical rescue is possible. This is the only place and time in our operational routine where it is acceptable to expose a worker to a level of risk that seriously challenges the personal capabilities of our firefighters (physical, mental, emotional) and the upper performance limits of our standard, operational safety system.

Identify Strategy/Develop Incident Action Plan

Safety Effect:

2. We will risk a little to protect savable property.

This is a situation where there is savable property that is threatened by the incident problem, and where the IC evaluates that the use of our regular, personal, and operational safety systems will adequately protect the workers, so that they can directly stabilize the incident problem, and protect that property. These are situations where the rescue priority has been completed, and the remaining level of risk is light to moderate. In these cases, routine offensive operations (within the capabilities of our safety system) can complete the incident stabilization function by protecting savable property.

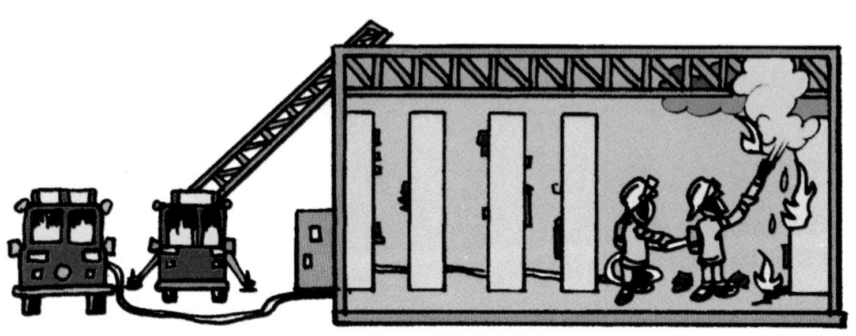

3. We will not take any risk to protect lives or property that are already lost.

This is a situation where the incident problem has evolved to such a point that lives and property are no longer savable. The IC must quickly evaluate and identify such conditions, and then (as a standard practice) declare and manage the incident (or that involved fire area) as a defensive operation. This is a critical time for our risk management plan and requires strong control of the position/function of hazard-zone workers by the IC and the command system, along with a disciplined response from our workers.

Command Safety

Identify Strategy/Develop Incident Action Plan

IC Checklist:

☐ Declare the strategy, as part of the IRR.

Safety Effect:

Having the initial-arriving IC declare the offensive/defensive strategy in the IRR begins the incident in a standard and clearly understandable operational inside or outside strategic mode. Declaring the strategy (vocally), as a regular part of the IRR, requires the IC to take the time, early in the event, that is required to (visually) evaluate conditions and to make a conscious (mental) decision about where the fire fight will occur, and where the fire fight will not occur, based on current and forecasted critical incident conditions. This requries the IC to translate the initial evaluation into a standard strategic category: offensive or defensive. This process makes the IC consider the critical strategic factors in a rational way and then declare (in standard operational terms: either offensive or defensive) how those factors add up in comparison to the capability of our standard safety system. We increase safety whenever a regular array of decision-making considerations can be used to create a standard response. Taking this time to evaluate and decide on the basic strategy, in the beginning of the operation, saves a lot of time (and pain) later on, because it correctly and safely positions our initial attack, and prevents us from

Command Safety

IDENTIFY STRATEGY/DEVELOP INCIDENT ACTION PLAN

Safety Effect :

having to recover from being in an incorrect (and unsafe) place--lots of times when we find ourselves in an unsafe spot, it's because in the very beginning of the incident, we acted instinctively (emotionally), and not rationally. When we make this basic initial ("emotional motion") position/action mistake, we create a ragged and screwed up (and many times unsafe) incident beginning. The initial confusion that emerges from being in an unsafe initial spot generally (caused by a lack of adequate evaluation) provides the front end for the classic unsafe "snowball" (effect) that continues to pick up "steam" (pun), if it is not somehow interrupted (by a strong IC), and can evolve (or devolve) into a full-blown cluster... the safety effectiveness factor goes down (exponentially) as the cluster factor goes up. Consistently having the right people in the right place, doing the right thing (i.e., performing the correct action for the current incident stage), is the best cluster prevention. A major safety factor in how we do our front-end work involves operating very quickly to solve a hazardous problem. Our operational speed can sometimes compete with an effective level of evaluation and decision making, particularly in the rushed, dangerous, and confusing initial stages. Requiring IC #1 to slow down (just a bit), evaluate what's going on, and report the beginning strategy in the IRR is an excellent investment in a safe, standard, sensible incident start. Taking a little time to do smart stuff in the beginning always sets the stage for correct action and saves time later on... to go faster--slow down.

Command Safety

Identify Strategy/Develop Incident Action Plan

IC Checklist:

☐ Manage and control operations, within the basic strategy.

Safety Effect:

The IC must create a standard operational response, after the offensive/defensive decision is made. Offensive fire fighting operations are conducted in the interior of the structure. Offensive conditions exist in a limited time frame and can quickly change (for the worse), if the incident problems are not controlled and eliminated. Based on the perishability of those offensive conditions, the inside attack should be fast, aggressive, and well placed. The offensive attack generally comes in through and protects the unburned side. The IC must attempt to control interior access by covering halls, stairs, lobbies, public places (where people normally move around inside the building). "Covering" means actually assigning (and holding) firefighters with tools (forcible entry, access, ventilation, lights, thermal imaging, and hose lines) in those places (physically), to keep the fire from extending into those places. When fire extends into

Safety Effect:

those "public" areas, we lose our inside access/movement options and the situation quickly starts to become defensive. This becomes particularly dangerous when an offensive attack is underway inside, where conditions change very quickly, and those public access places become involved. When this happens, the fire is, in effect, removing the exits... this can be a big deal (particularly for those who are inside). The IC must estimate how long fire/structural conditions will allow firefighters to operate on the inside, and then control operations within that time frame. This means (simply) when the offensive time is up, if conditions are not getting better, the IC must order the insiders out. In these deteriorating situations, it's (lots) better for the IC to anticipate that defensive conditions are on the way, and to move the troops ahead of the arrival of those conditions, rather than having the fire (literally) chase the firefighters out of the interior, or worse yet, having firefighters trapped in offensive positions by defensive conditions that moved in "behind" them.

Defensive fire fighting operations are conducted on the outside and safely away from the involved building (i.e., outside the collapse zone), when conditions and hazards are so extensive/severe that firefighters cannot safely operate on the inside. In these cases, the IC must clearly establish an outside, defensive operation. Defensive efforts should surround the fire area, and protect exposures with large, ground-level and elevated fire streams. Operational sector and safety officers should

Command Safety

Identify Strategy/Develop Incident Action Plan

Safety Effect:

quickly establish and tape off collapse zones. Large defensive jobs are many times long-term exercises (as opposed to typically quicker, offensive operations) that involve letting the fire burn the fuel down to the extinguishing capability of the water application (BTUs vs GPMs). During defensive operations, everyone must stay in safe positions outside the collapse zone, eat some Snickers candy bars in rehab, and patiently conduct what is essentially a water show. Remember, defensive operations mean you are outside the building, applying water to the inside. When this occurs, it is not sort of defensive--it is defensive.

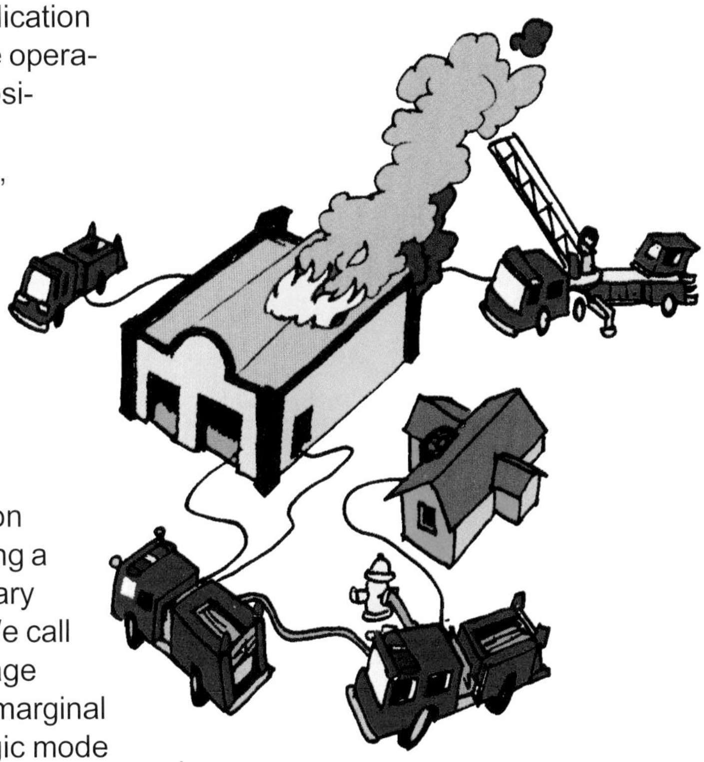

Sometimes, the IC will hold troops on the inside (longer than normal) during a later-offensive stage when the primary search has not been completed. We call this rescue-oriented operational stage "marginal." Everyone must realize marginal is a stage and not a regular, strategic mode (in spite of what some of our thrill seekers believe)--the only reason we should still be inside is to finish the rescue priority. These life-safety operations are always dangerous, and are the legitimate time (and reason) where we risk our lives a lot to protect a savable life--this is part of our standard, risk management plan. During marginal operations, the IC must carefully control the position and function of the rescuers. This means maintaining strong communications, protecting the insiders with attack and backup lines, having rescue (RIC) teams ready and being

Safety Effect:

pessimistic about ordering them out of the hot zone, when necessary. Giving a specific radio report helps to describe a marginal situation, i.e., "Ajax Command, we are in the offensive strategy working under marginal conditions in the unburned area, attempting to get an all clear."

The IC must always separate inside/outside operations from occurring in the same place. Maintaining and managing this offensive/defensive separation is a major way the IC protects the troops. If crews are working inside a structure, the IC must control exterior operations, and stop outside units from blasting the insiders. Separating offensive/defensive strategies keeps us from killing each other. If the IC makes the strategic operating position decision, based on the critical factors and the risk management plan, that an offensive strategy is no longer a viable option, then the IC must change the attack to defensive. This keeps the structure and the products of combustion from killing the firefighters. Maintaining strategic control is like the IC operating a traffic signal--the light is always automatically blinking red on our arrival. This requires everyone to stop, until the IC decides on the strategy, activates the strategy light by turning on and continuously maintaining only one color--green, yellow or red:

- ☐ green = offensive
- ☐ yellow = marginal
- ☐ red = defensive

Command Safety

IDENTIFY STRATEGY/DEVELOP INCIDENT ACTION PLAN

Safety Effect:

Remember that marginal is a stage (the end of offensive), not a regular strategic mode. The IC must evaluate conditions and the effectiveness of operational action, as it affects firefighter safety and then maintain the color that goes with that strategy decision. Like any traffic signal, the system is clearly programmed, so that only one light (color) can be on at a time--simply, the colors don't mix, and unlike a regular (intersection) traffic signal, our strategy light must be the same color on all sides (inside, top, bottom, and four sides). The objective of the signal is to create a standard action that goes with a standard condition, in order to achieve a standard outcome--the IC must always be the "keeper of the light."

The IC must remember that nobody, except us, wears turnout gear and SCBAs, and they must know (and react to) the difference between a rescue and a body recovery, and continually apply a street-smart level of reality to marginal operations. The marginal mode is designed to extend the period of inside operations, in well-evaluated and controlled situations where firefighters can go to the limits of the regular safety system, to take a chance on rescuing what is evaluated to be a viable customer trapped in the hazard zone, who the IC evaluates has a chance of being rescued. Superheated areas, charged with smoke, are absolutely (and conclusively) lethal, and those conditions kill unprotected humans, almost instantly. Because of the natural attack inclination of our troops and modern PPE capabilities,

Command Safety

Identify Strategy/Develop Incident Action Plan

Safety Effect:

sometimes firefighters penetrate (deeper/longer) hazard areas where they shouldn't. Marginal operations must be the product of a well-thought-out IAP. The marginal decision must be based on reliable, realistic (survival) information. The IC must evaluate the survivability profile of unprotected customers pessimistically, and not be distracted by the old time "we thought someone might be inside" free-lance reason for being in an unsafe interior position (this is known as "The Rescue Alibi"). In these cases, the IC must cause the troops to stay out, or pull out of the hazard zone. Marginal times are just that--marginal, and must be carefully, quickly, and critically managed. The IC must be pessimistic when dealing with marginal situations--they generally involve conditions that are rapidly changing, and not getting better. If not quickly brought under control, these conditions can (and soon will) overcome the protective capability of even our standard, safety system--so imagine how long poor old Fred Smith will last (who was dumb enough to actually forget to put on his SCBA and designer nomex sleep ensemble, and then went nighty night, dressed only in his polyester pajamas).

Command Safety

IDENTIFY STRATEGY/DEVELOP INCIDENT ACTION PLAN

IC Checklist:

☐ Implement an IAP to match the overall strategy.

Safety Effect:

A major part of incident command and control involves the IC first deciding on the overall strategy, and then developing a matching IAP. First the initial size up, then strategy, then IAP, then the ongoing size up to stay updated... in that order. The standard strategy decision is where we pick either inside/outside, and that provides the (overall strategic) positional basis for safe and effective operations. The IAP is where the strategy is translated into tactical terms, so that it can "go to work." The IAP provides the operational details, within the selected strategy, that achieve the standard tactical objectives, that actually describe and then create the manual labor that addresses and solves the incident problem.

Standard Tactical Objectives:

1.	firefighter safety	... major focus throughout operations
2.	rescue	... the operational work we do while there is a hazard zone present
3.	fire control	
4.	loss control	
5.	customer stabilization	... we don't leave until the customer is connected to short-term recovery resources and services.

Command Safety

IDENTIFY STRATEGY/DEVELOP INCIDENT ACTION PLAN

Safety Effect:

Standardizing these tactical activities, along with the benchmark of completion for each, eliminates a guessing game at show time. This standard priority-based approach creates a more predictable, dependable, and safer incident operation because everyone on the team understands, and can depend on the order of our tactical efforts.

(this is what we do and the order we do it in)	(this is what we say when we finish so we know we are done with this activity and can go onto the next one)	(this is what it looks like when we do it)
STANDARD TACTICAL OBJECTIVES ⇨	STANDARD BENCHMARKS OF COMPLETION	⇨ WHAT BENCHMARK MEANS
1. firefighter safety	PAR given incrementally at standard times	crew is together in their assigned position and okay
2. rescue	"all clear"	primary search is complete
3. fire control	"under control"	forward fire progress has been stopped and the on-scene resources can complete extinguishment
4. loss control	"loss stopped"	all damage has been stopped and all remaining property is protected
5. customer stabilization	"customer okay"	the customer's life has been stabilized and we have provided short-term resources to them to assist them with getting to the next stage of recovery

Command Safety

Identify Strategy/Develop Incident Action Plan

Safety Effect:

The details of the IAP are structured around the basic, five tactical priorities, and are what goes into the "boxes" on the IC's TWS. Performing the incident size up and making operational decisions is how the IC fills in those boxes, to produce the specific orders to specific companies, in specific places, that accomplish specific tasks. Together the separate (but hopefully integrated) tasks achieve the basic tactical objectives that create an effective, overall problem-solving effort. The IC must develop and implement an ongoing array of orders to working units/ sectors that "pull off" the entire IAP. It is critical that the IAP is based on a conscious, deliberate development of the incident strategy, and not an automatic hazard-zone attack based on an emotional reaction to obvious (fire-showing) incident conditions. Having the IC consider a full range of the standard tactical options creates a balanced evaluation of what is going on. A smart, initial IAP, that is based on a complete evaluation, prevents attackers from having to mutter, "What the hell are we doing here," after they get seduced into an unsafe position by a "candle-moth" sucker punch.

Command Safety

Identify Strategy/Develop Incident Action Plan

Note:

In this section, five tactical priorities have been listed--in other places, we have used only rescue/fire control/loss stopped priorities, because they are the basic operational activities we do (and the order in which we do them), while there is a hazard zone present. This entire blabfest is devoted to presenting the IC's role and responsibilities to effectively protect the hazard-zone workers. So in effect, everything you're reading is directed toward the #1 priority "firefighter safety." Safety procedures must be performed throughout the incident operation--from our initial response, while we are doing operational action, and during the return trip back to quarters. Simply, the "safety light" must always be engaged on the strategic, tactical, and task level. Customer stabilization is absolutely critical, and is an activity to which we are all now paying more attention. How we preform at the end of the event, and how we "leave the customer," creates a huge part of the memory they have of their experience with us. This customer-centered activity typically occurs after we have dealt with, and hopefully, eliminated the hazard zone, so it has not been a major part of this manual. We have tried to focus on just the period when there is a hazard zone present and this, in no way, is meant to minimize how important our treatment of the Smith family is to our service (please see *Essentials of Customer Service* which is available through the International Fire Service Training Association [IFSTA]).

Command Safety

Identify Strategy/Develop Incident Action Plan

IC Checklist:

☐ Use critical factors to develop the IAP.

Safety Effect:

The IC must use the same critical factors (presented earlier in the biblical length Command Safety Function #2, "Situation Evaluation") for developing the overall strategy, the IAP, and an evaluation (and reaction) to how well the IAP solved the problems, caused by the critical factors. The approach of using the same factors (for all three evaluation purposes; strategy/IAP/outcome) creates a multiple-duty "mental overlay," where the factors become familiar, and over time, the experiences the IC develops in dealing with them get loaded into, and increase the mental files on their dynamics. Connecting and integrating the strategy/IAP produces an action-oriented outcome, that emerges out of the actual orders that produce problem-solving, manual labor (the reason we respond). Effectively "adding up" an evaluation of how the factors look becomes the basis for the IAP, that provides for the safety and welfare of the workers. The IAP is a short, simple "statement" of the work plan that is required to

Command Safety

Identify Strategy/Develop Incident Action Plan

Safety Effect:

complete the tactical priorities for the incident. This is also where the IC turns the evaluation/decisions about the critical factors into a practical plan. This is where the IC actually applies the standard risk management plan, and that plan becomes an important part of the orders the IC extends to the responders, who will act out the IAP. This is where the IAP (and the IC) "goes to work"... for the IC and the entire response team, the IAP is "where the rubber meets the road." The IC uses an evaluation of the conditions (factors) to produce the overall incident strategy. Then this gets translated into some type of lucid action plan that is based on both the overall strategy and the critical factors. This leads to the tactical assignment of companies and other resources. When this is done correctly, it neatly ties what is happening (conditions) to what needs to be done (action), how it is going to get done (orders), and who is going to do it (assignments). It provides the IC with a system to evaluate and decide how service will be provided to the customer, and how to manage the incident so no one gets hurt.

Going from the initial evaluation of critical conditions to actual orders to companies and sectors must occur very quickly (in the street). Experienced ICs develop a regular mental method that quickly processes the evaluate/decide/order steps almost automatically. This (automatic) ability requires the IC to quickly connect a fairly, standard operational response to a safety/rescue/fire

Command Safety

Identify Strategy/Develop Incident Action Plan

Safety Effect:

control/property conservation/customer assistance problem. This requires the IC to understand what it takes to solve a particular "piece" of the incident, and then assign that task to a team of workers. The IC must divide the incident up into a set of doable tasks, that cover the geographic and functional needs of the whole incident. The basis of doing this is knowing how much standard work a standard unit can perform, and staying within the capability/limitation of that performance framework. This becomes a huge safety factor, simply because it prevents the IC from being unrealistic (and unsafe), when they assign work to workers. Attempting to describe the process (in written words) becomes complicated and awkward--it's sort of like trying to describe a kiss (if it's a good one). Perhaps the diagram on the next page helps--sorry about the inept, older author (did not pay attention in English class because he was fantasizing about kissing).

Command Safety

Identify Strategy/Develop Incident Action Plan

IAP DIAGRAM

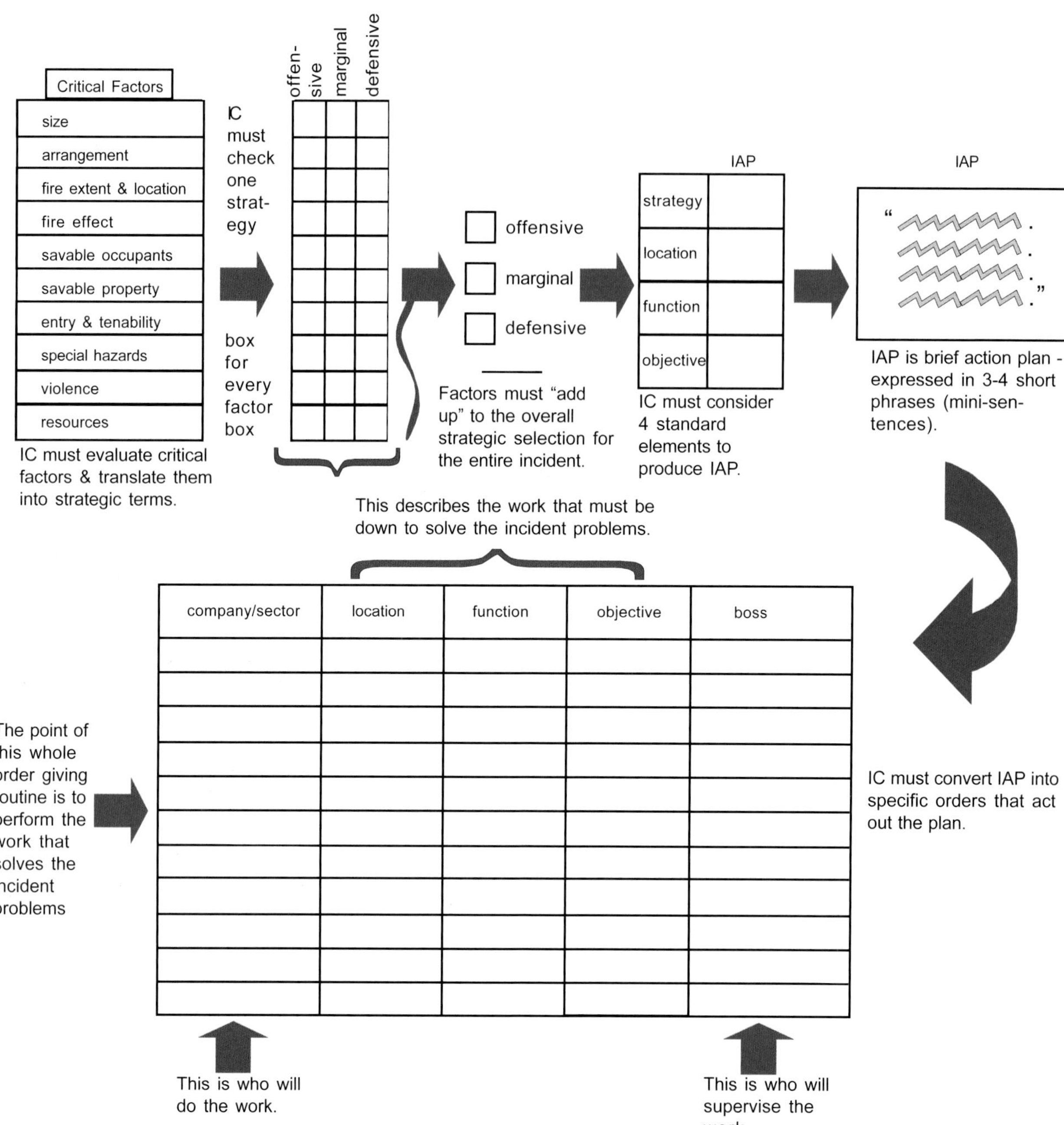

Note:

All of these boxes and arrows create a fairly simple conceptual framework (flow chart) that can be applied to self study/classroom/simulator training situations. However, all the boxes, arrows, and squigglies probably go beyond our normal human ability to use it under actual incident conditions. Its value is in providing a framework that we can use to increase our IAP development understanding and for command practice, so it gets loaded (conceptually) into the IC's noggin' so they can develop the skill habit and routine ability to quickly and comfortably apply it at an incident. The TWS becomes a simple version of this approach and is designed to be simple enough to be used by the IC under actual incident conditions.

Command Safety

Identify Strategy/Develop Incident Action Plan

IC Checklist:

☐ Include strategy, location, function, and objective in IAP.

Safety Effect:

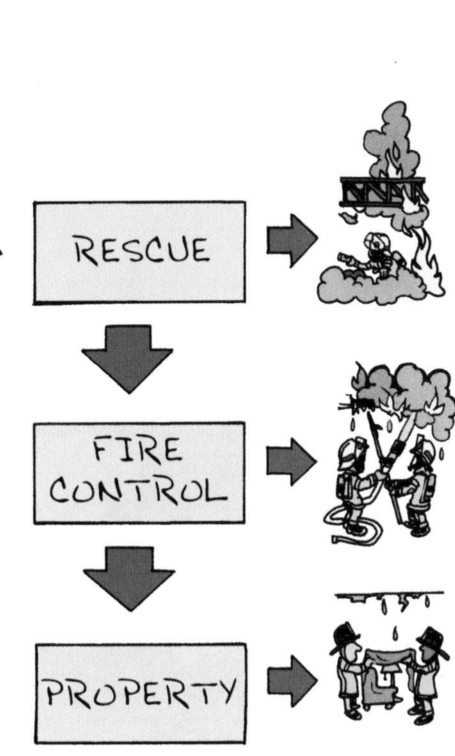

The IAP becomes a short, simple expression of the IC's plan to solve the incident problem. It is the answer when the IC's boss (or really anyone) asks, "What's your plan?" It must be a brief, accurate description of the where, what, and who of operations, and should include:

☐ Strategy

The IC declares the strategy in the IRR and this then becomes the basis of operations. Responders can both hear that strategic declaration over the tactical radio frequency, and can personally evaluate conditions themselves upon their arrival. It certainly makes a lot more sense to those responders (or really anyone) if their view (and evaluation) of incident conditions match the IC's strategic report. Having the IC assign units to the operating positions and functions that act out the IAP (after the IC has declared the overall strategy) is another standard place where the troops get to evaluate if the declaration/visualization/orders all strategically "line up" and make sense.

Command Safety

Identify Strategy/Develop Incident Action Plan

Safety Effect:

☐ Location

The IC must develop a plan for the geography (and geometry) of the incident site in relation to the conditions that are present. This creates a location-based work plan for the critical places the operation must physically cover. The IC must develop a priority list of the order in which basic operational and command positions must be covered. It is important that the critical places be identified and effectively named, particularly as it relates to in/out access. It becomes a major safety factor to identify and use standard names to describe building sides (A, B, C, D), doors, windows, floors/levels, building openings, even directions (North, South, East, West) become a critical safety factor. Lots of firefighters have gotten scuffed up (and lots worse) because (as an example) they thought they were on the first floor (in the front), and the troops (in the rear) say they are on the third floor, because the building is on a hill.

☐ Function

The IC must develop a brief description of the basic function that will be performed to solve the incident problem. If there are multiple locations where major efforts will be extended, a function must be connected with coordinated supervision to each of those separate spots. Particular coordination is required when

Command Safety

Identify Strategy/Develop Incident Action Plan

Safety Effect:

the separate functions "bump into" each other (or "splash" into each other with fire streams), in a way that can affect the safety of the workers. The functions must always fit into and match the overall strategy/IAP.

☐ Objective

The IC must describe the basic objective of the overall operational effort. This IAP component becomes an important description of what the plan is trying to achieve. The basic tactical priorities and their benchmarks of completion become the framework for these objectives. While this sounds very simple, it becomes a major safety factor that the IC has evaluated conditions and has developed a conscious objective upon which to base the IAP. The command team must continually focus on the overall operational objective to be certain the IAP is doing the right things, in the correct places.

Identify Strategy/Develop Incident Action Plan

Safety Effect:

- ☐ Use tactical priority benchmarks, as the IAP road map.

Safety Effect:

Structural fire fighting operations revolve around three basic tactical priorities: #1 Rescue; #2 Fire Control; #3 Property Conservation/Loss Control (safety and customer assistance must occur throughout the operational and recovery period). These priorities describe the action we will take during the operational period when there is a hazard zone present, and this list of activities provides a simple, basic game plan that creates a predictable, understandable team approach. Knowing what our tactical efforts will be directed toward, and the order of those efforts, increases incident safety because the IC can attach a different risk level to each priority. Life safety is where we will take the biggest (legitimately permitted) risk. It is the first objective the IC attempts to complete, by extending a primary search in the unburned areas of the fire building and exposures, where our standard safety system will protect the firefighters. Search order:

- ☐ most severely threatened
- ☐ remainder of fire area
- ☐ rest of building

When the primary search has been completed and an "all clear" is transmitted (benchmark of completion), the IC is now

Safety Effect:

protecting property--not people. Now the IC must reassess the situation, and apply the medium, acceptable risk level that is required to continue offensive operations, to what are now fire fighting efforts. During fire control, the IC must continually evaluate the ongoing relationship between the level of hazard present, against the capability of our regular safety system. When the fire has been controlled, and the "under control" has been transmitted (second benchmark), the IC must again shift safety gears and go to a no-risk level of operation. We now have the luxury of time to light up, ventilate, and carefully evaluate what additional property conservation (loss control) work must be done.

Firefighters can now go to rehab, get out of their PPE, cool off, hydrate, Snicker bar up, and, if necessary, be evaluated by medics. During property conservation/loss control phases, the main incident problems have been controlled, so the hazards are residual and typically less active. Many times, they (the hazards) include operating in and around a structure that the fire has beaten up and weakened, very dangerous (but typically less visible) residual (but still hazardous) products of combustion, tired firefighters, exposure, cuts, bruises, and the manual-labor scuffs that occur during overhaul. Rotating fresh crews in and maintaining strong, ongoing operational/safety sector support to complete overhaul is a smart approach. The standard tactical priorities create a sensible, predictable, dependable

Command Safety

Identify Strategy/Develop Incident Action Plan

Safety Effect:

game plan for how our work will occur, and creates the framework for the safety routine/risk management that must be applied to each separate priority. We are always safer when we follow a standard, safety-based operational approach we all understand and agree to ahead of the incident.

Note:

The IAP is based on the strategy and critical incident factors. The critical factors should also drive the set of tactical activities the IC does (IAP) to achieve the regular tactical priorities. Using this formula will sometimes cause the IC to do fire control first, to buy the needed time to safely and effectively complete search and rescue (example--a controllable fire in a large building that is quickly filling up with smoke... a single, localized fire in a hotel that is extending into the common hallway).

Command Safety

Identify Strategy/Develop Incident Action Plan

IC Checklist:

- ☐ Reannounce ongoing strategy confirmation, as part of ET reports.

Safety Effect:

A regular part of the IC's evaluation process should be periodically restating the strategy, over the tactical channel at regular intervals. This forces the IC to reevaluate the critical factors throughout the fire fight. Specifically declaring the current overall strategy over the tactical channel keeps everyone informed about where they should be, and what they should be doing (and where they should not be, and what they should not be doing).

One of the ways the IC creates a safe operation is by building an incident organization that covers all the critical incident places and functions. The decentralized organization must evaluate critical factors in their area, and then "push" information on those conditions to the IC. The IC receives and processes that intelligence coming into the CP, and must periodically reaffirm the strategy. The standard, interval-based, ET reporting process keeps the IC/command team on track, and "strategically honest' (it's tough to have to "fib" about the strategy being offensive every five minutes). Another advantage of the periodic reporting of the strategy is that it requires the command team to logically connect elapsed-operational time with the progress (or lack of) that has occurred, and to make that connection in a

Safety Effect:

way that can be acted out in a standard way. Declaring offensive or defensive is the shortest, quickest, and most powerful way for the IC to say something that can be understood and reacted to, by virtually everyone at the incident. We know how long typical offensive conditions will last and how long the completion of tactical priorities takes. Making standard reports, at prearranged ETs, creates the basis for command players to quietly listen and mentally nod their heads, as work gets completed "on time" and reported, and where visual information verifies that the operation is progressing normally. Just the opposite is also true. Developing and using a standard way to describe standard action, and then attaching a reporting-time frame to that action creates a certain incident "logic" for the IC. As an example, when someone in the command team hears thirty-minute ET/still in the offensive mode/we haven't got an "all clear" and haven't started getting the fire wet (yet), unless something very exotic(!) is going on, they are going to get a PAR and do a major reevaluation (and possible regroup). Standardizing all these critical pieces of the system creates a regular pattern that the IC uses to evaluate and effectively react to as the incident evolves. When the incident pieces fit into the regular pattern, the IC can quietly proceed with the IAP. When some standard action doesn't create a standard outcome, it becomes a major "red flag" for the IC, that requires a quick and sometimes active (not quiet) command response. Understanding and managing those patterns

Safety Effect:

become a major safety deal. Many times such nonstandard responses mean that the condition they were applied to was indeed not "standard." These conditions, that are not accurately identified, can cause big-time safety problems, and generally require quick, strong action to control. The effective response to those patterns emerges out of SOPs, study, reflection, road rash, and refinement. This is why we say, "Give me a new Suburban and an old IC (in wisdom, not age).

Command Safety

Identify Strategy/Develop Incident Action Plan

IC Checklist:

- ☐ Do not combine offensive/defensive operations in the same fire area (compartment).

Safety Effect:

This is very simple. The IC must not let anyone on the outside put fire streams into the same area where crews are operating in the interior. Do not do this... ever. Enough said.

Command Safety

Identify Strategy/Develop Incident Action Plan

IC Checklist:

- ☐ Use the incident organization and communications to connect and act out strategy/IAP.

Safety Effect:

The IC must use a regular system that effectively develops and manages the strategy and the IAP, and provides two-way communication on the ongoing status of both of them. Lots of surprising, unsafe, sometimes fatal outcomes occur when critical information is not distributed throughout the incident operation--particularly in (and out of) the hazard zone. Creating this on-line, two-way information system under actual incident conditions is a difficult process, because things can change quickly in one spot, that can dramatically affect the safety of the entire incident operation. Having a fast, agile, responsive reporting capability supported by an effective incident organization, that is automatically put in place at every event, can make the difference between life and death. The organization must empower (actually require) every member, company, and sector to quickly evaluate and report up, across, and down critical conditions they see from their position. This safety information becomes the highest priority radio air time. In very urgent situations that involve the welfare of the workers, emergency traffic should be used to clear the tactical frequency of routine radio traffic. The IC must become "critical information central" (focal point), and coordinate the overall reaction to those conditions. Lots of

Command Safety

Identify Strategy/Develop Incident Action Plan

Safety Effect:

sad firefighter injury/death events were caused by conditions, someone knew about and kept a secret, until it was too late--it's better to sensibly overcommunicate, and overreact to safety stuff, than the opposite. Operating companies must react appropriately to the actual conditions they encounter. If the IC declared the strategy as offensive, and assigned a company to a position (offensive) and they encountered defensive conditions, they must react accordingly. This means that the company withdraws to a safe position (without asking for permission), and then notifies the IC of their action (and the defensive conditions) over the radio.

Command Safety

Identify Strategy/Develop Incident Action Plan

This happens when Strategy/Incident Action Planning (IAP) is done:

- ☐ Strategy agreement (IAP) provides basic position/function plan ➡ we start/stay standard, safe, and smart.
- ☐ Strategy drives IAP (not the opposite).
- ☐ Strategy is always connected to the basic risk management plan:
  - risk a lot ➡ protect savable life
  - risk a little ➡ protect savable property
  - no risk ➡ lives/property already lost.
- ☐ IC assigns resources in correct positions to achieve standard strategic objectives.
- ☐ Rescue/fire control/property conservation get completed in correct priority order.
- ☐ IC reevaluates risk level at the end of each tactical priority and readjusts strategy and IAP to match risk management plan.
- ☐ Firefighters are in offensive positions at offensive fires, and in defensive positions at defensive fires.
- ☐ We don't mix offensive/defensive in same place, at the same time.
- ☐ Troops: safe--survive (we don't die for property).
- ☐ Standard conditions end in standard outcomes:
  - offensive--fast, inside, knock down, effective exit
  - defensive--surround, drown, fire area gonzo, protect exposures; sometimes we save the building (or part of it) with a well-placed defensive effort.
- ☐ Allows IC to stay ahead of "power curve."
- ☐ We have IAP (+ strategy) to critique... we learn and "file" lessons.
- ☐ IC thanks troops for effective action/outcome.

Command Safety

Identify Strategy/Develop Incident Action Plan

This happens when Strategy/Incident Action Planning (IAP) is not done:

- ☐ Firefighters end up in offensive positions at defensive fires (very bad).
- ☐ We don't base ops on conscious strategic decisions = strategic free-lancing.
- ☐ No position/function control--lots of incorrect, conflicting, painful positioning--we play catch up with Mrs. Smith's anatomy and physiology, and her stuff.
- ☐ Incident-action driven by free-lancing--no IAP plan.
- ☐ We have no standard risk management--free-lancers do lots of stupid stuff--no focus on firefighter safety--we take big risk for no gain.
- ☐ We have lots of offensive/defensive mixture in same place... very dangerous.
- ☐ Companies reinforce bad positions, submit to "candle-moth" tunnel vision.
- ☐ Action is reactive not planned--lots of confusion = lack of strategic control is the basis of confused, unsafe operations.
- ☐ No learning--hard to critique non-plan (cluster).
- ☐ We get to wear uncomfortable wool dress uniforms, and listen to very sad bagpipe music (big funeral ritual).

Command Safety

Identify Strategy/Develop Incident Action Plan

COACHING VERSION

"Quickly and consciously decide if we go inside or stay outside--wherever possible, front-end load a rapid, strong, offensive, inside attack, but control those overly aggressive souls who have a post-hypnotic suggestion that all initial operations are automatically offensive, and instinctively jump into the interior, even if they are about to wear it (Kamikaze fire fighting). Always evaluate our ability to make a round trip into and out of the hazard zone. Never forget that a safe exit is a critical part of every offensive entry--be careful of situations that will allow us to get in, but then will not let us out. Our strength is quick force, particularly in the beginning--if you aren't bigger and badder than the problem--the problem wins... so stay out of its way while it does. Use the strategy/IAP incident organization and effective commo to stay connected to the troops, and always be ready to change strategy/location to match changing conditions. Pay attention to the basic risk management plan and do what it says to do... never, ever, trade live firefighters for dead customers, or for any piece of involved or exposed property. Initially and periodically, ask yourself, "Is what we're trying to accomplish worth the risk to our human resources (firefighters)?" When you get an "all clear," reevaluate and determine the current risk level. If there is a small risk, keep going on the inside/if there is a big risk, you must go defensive. Remember, you didn't light the fire--God meant some buildings to burn (basically, the unsprinklered ones).

Command Safety

IDENTIFY STRATEGY/DEVELOP INCIDENT ACTION PLAN

Command Safety

Build an Incident Organization

CHAPTER 6

Command Safety

Build an Incident Organization

Command Safety

Build an Incident Organization

Function 6

Build an Incident Organization

Major Goal

To develop an effective incident organization using the sector system to decentralize and delegate geographic and functional responsibility.

IC Checklist:

- ☐ Quickly develop an incident organization to keep everyone connected.
- ☐ Match and balance organization to size/structure/complexity of the operating resources.
- ☐ Forecast and establish geographic/functional sectors.
- ☐ Accomplish effective delegation and span-of-control management through early sectoring.
- ☐ Correctly name sectors and landmarks.
- ☐ Assign and brief sector officers.
- ☐ Limit units assigned to sectors to five (5)
- ☐ Serve as resource allocator to sectors.
- ☐ Build a command team.
- ☐ Build outside agency liaison into the organization.
- ☐ Operate on the strategic level--support tactical and task levels.
- ☐ Evaluate progress reports, assist, and coordinate sector activities.
- ☐ Implement management sections and branches to support operational and command escalation.
- ☐ Use organization chart as communications flow plan.
- ☐ Allow yourself to be supported in the process.

Command Safety

Build an Incident Organization

IC Checklist:

☐ Quickly develop an incident organization to keep everyone connected.

Safety Effect:

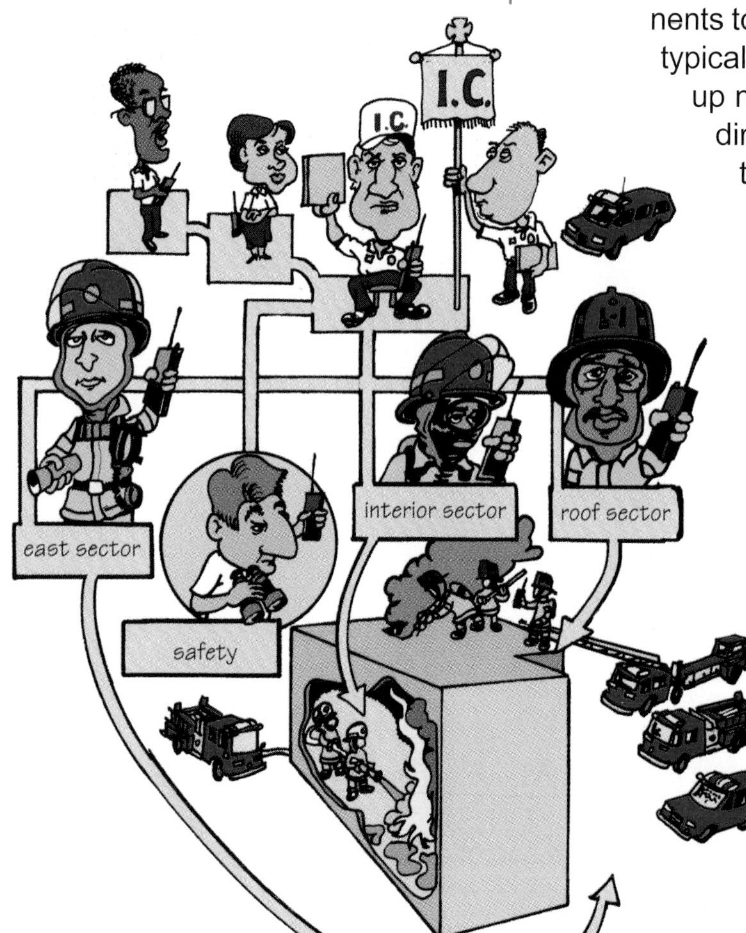

A major way the IC maintains control of the incident is by using standard IMS components to build an incident organization. A typical local incident organization is made up mostly of company officers, who directly supervise their crew, and sector officers, who serve as tactical-level bosses in key tactical positions and in major functions. Sector officers work for/report to the IC, and company officers work for/report to sectors. Geographic sectors manage a portion (area) of the fire building or fire area. Functional sectors are responsible for a support function and work mostly on just that single (their assigned) functional activity. Sector officers actually (physically) go where they are assigned, and directly supervise the task-level work teams in that area. The sector system becomes a principal way the IC maintains an effective and manageable span of control. The IC delegates work assignments and shares (i.e., "spreads out") the responsibility and authority required to create effective bosses all over the incident. Managing the safety and welfare of the troops is

Build an Incident Organization

Safety Effect:

a major function of sector bosses. They do this by directly evaluating conditions and managing the adjustments and changes in their work area. They use their decentralized "close to the physical work" location to manage and move the troops based on what is happening, and what they forecast will happen. They become direct safety and operational managers and reporting agents (see/hear/sense/report) for the IC, who is in the non-direct see/hear/sense position inside the CP. From that CP position, the IC generally cannot see (at least not in much detail) what's going on in the sectors (particularly inside/opposite side). Sectors are responsible for maintaining an ongoing awareness (see/hear/sense/report) of the who, where, what, when (time) of the resources assigned to their sector, along with conditions in that area/function, work progress, and the ongoing safety and welfare of the workers assigned to that sector. Managing these details directly within the sector creates an effective level of centralized (IC in CP) and decentralized (sector) management, and safety all over the incident site. The IC maintains an ongoing centralized (overall) awareness of incident conditions because the sectors report that decentralized (area/function) information back to command. Having these up close and personal company/sector "eyes and ears" all over the incident is the major reason the IC can stay put in the CP. The IC uses the CP advantage to receive, assemble, process, and react to input from all the sectors, and decides how that processed information fits

Command Safety

Build an Incident Organization

Safety Effect:

together, as everyone works on safely completing the IAP. The material in this section describes the very practical difference between the strategic level of the IC and the tactical level of sector operation. The actual existence of both the IC and the sectors (operating together, each on their own level) creates the street-level capability for each to perform their standard role and function--without sectors operating in tactical positions, seeing, hearing, sensing and reporting, the IC can't set up, operate, and stay in an effective strategic position... without the IC in that CP position, the sectors don't have anybody to report to, who is anywhere close to being in an effective position to receive/process/react to what the sector is seeing, hearing, sensing, and reporting on their assigned location/function, or the combination of what all the sectors are seeing, hearing, sensing, and reporting.

Command Safety

Build an Incident Organization

IC Checklist:

☐ Match and balance organization to size/ structure/complexity of the operating resources.

Safety Effect:

The size, severity, duration, dynamics, and nature (complexity) of the incident will define what type of organizational pieces the IC puts into place. Big incidents require big organizational structures. Little incidents need small organizational structures. Highly hazardous incidents (regardless of size) require very careful (almost micro) management. A major organizational function is to achieve an effective balance between bosses and workers. Given the need to basically use manual labor to solve the incident problem, a good rule of thumb is the IC should have more people working than commanding. While the IC uses the same basic IMS elements to manage every sort of incident, EMS, haz mat, or special operations, the incidents will just have a different set of sectors (i.e., different sector names and the activity that goes with that name) than a structural fire. Matching the organization to the size, scope, nature, and stage of the incident provides enough of the right kind of

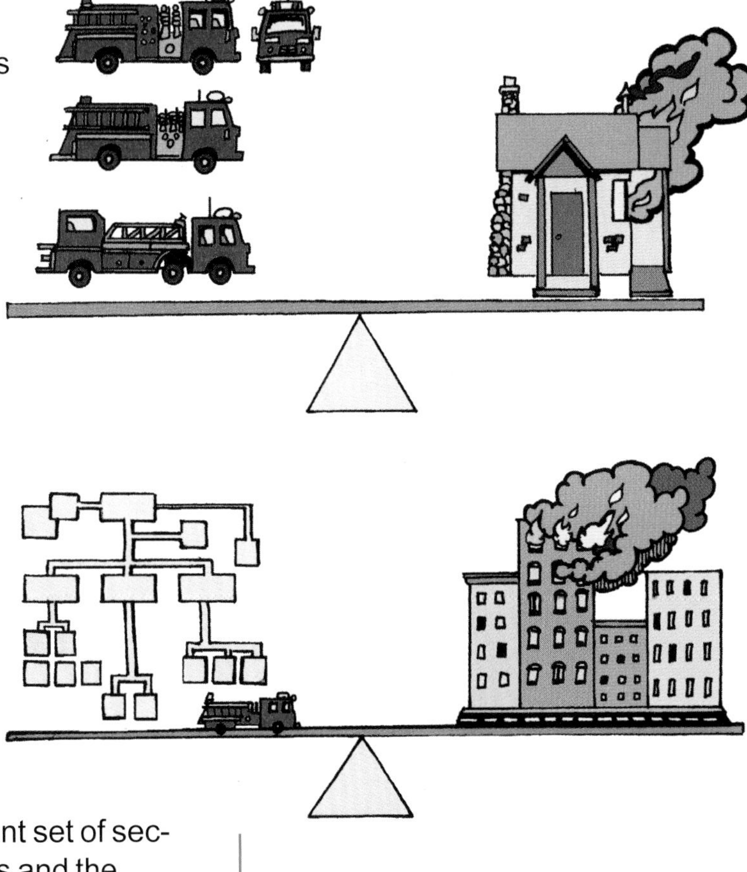

Command Safety

Build an Incident Organization

Safety Effect:

management and supervision to get the job done as effectively and safely as possible. The longer the command and operational team builds and uses incident organizations, the more routine and dependable the process becomes. Everyone (who understands the local command system) should be able to look at an incident and accurately predict the basic geographic/functional sectors that will be used to cover what needs to be done. They should be able to estimate how many companies will be assigned to an operational sector to do the work that must be accomplished, based on the conditions in that area/function. Another important command system capability that is developed over time is how the sectors interact with each other in a complimentary (synergistic) way. It requires skill and experience to effectively manage how the "edges" of geographic sectors touch and integrate with each other. This organizational integration becomes a critical safety (and operational effectiveness) consideration, because sectors can (as an example) either help each other stay in the structure/fire area and get the job done, or blast (literally) each other out of the building. The IC must look at the whole incident as a puzzle, and then make the sectors the individual pieces that "overlay" and cover the entire puzzle. Creating complete sector coverage is a major part of what it takes to create a safe and effective outcome. Sectors must also coordinate and manage the accountability of those assigned resources. When sectors "overlap," it is easy for a company to bounce back and forth between each sector. This can be a problem

Build an Incident Organization

Safety Effect:

when the sector officer needs to get a PAR for all their companies, and comes up PAR short. Sectors that operate in close proximity to one another should maintain a commo connection to conduct periodic status and roll calls with one another.

Note:

The IC builds the organization to safely and effectively manage the work that must be done to solve the incident problem(s). Organizational development is not an exercise in creating some elegant management design to fill some predetermined number of "boxes" to satisfy the ICS gods, or to put a body inside every sector vest stored in the trunk of the shift commander's Crown Victoria. Complicated organizational designs (or sizes) are impossible for responders to mentally understand, and difficult to physically (and managerially) pull off. The IC must assume a "simple-is-better" approach (to most things). Many times in quickly resolved, small- to medium-sized (local) offensive situations, the IC will solve the incident problem by directly managing the companies on the initial response. In these cases, companies will be assigned to key tactical positions and generally, their

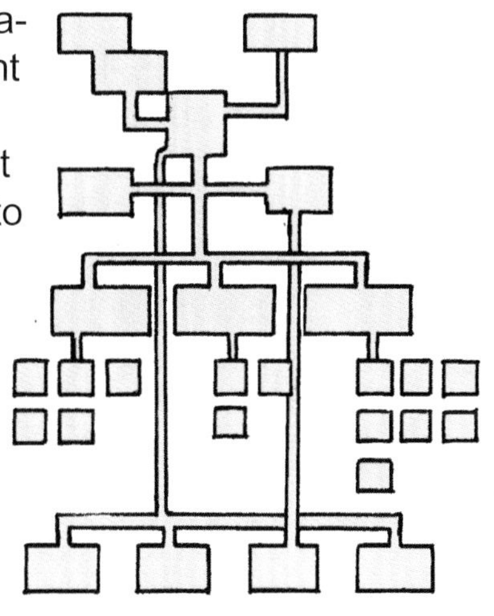

Command Safety

Build an Incident Organization

Safety Effect:

effective efforts will quickly solve the problem in that area. When this happens, the IC holds the assignment, stops building the incident organization, and the operating companies put out the fire, clean up, and conclude operations. In these cases, the IC really doesn't need to say very much beyond making (and recording on a TWS) the initial key assignments and starting a very simple (one person) command organization. If the place gets searched and the fire goes out, the IC can quietly watch the work being done, and coordinate whatever needs coordinating. In fact, the best command jobs are the quietest ones, and the best ICs are the ones who set up and manage the incident so they never have to raise their voices. If the incident continues to escalate, then the IC uses the initial (quiet) command foundation as the basis to expand the organization to get ahead of the problem. A big part of the "art of command" is being able to develop just the right amount of management/organization to be effective and safe. Too little is dangerous and exposes firefighters to unmanaged hazards. Too much organization restricts effective action and fails to take advantage of the spirit and capability of our humans.

Command Safety

Build an Incident Organization

IC Checklist:

- ☐ Forecast and establish geographic/functional sectors.

Safety Effect:

The IC should consider assigning sector officer responsibilities to the first company officer who arrives in key tactical positions. This early sectorization becomes particularly important if the IC forecasts that, based on the work that must be done, more resources will be required in those areas. If the first company solves the problem quickly, the problem goes away and the organization has no need to get bigger in that area. If the problem continues, the sector system that has been started early in the operation is already in place, and more resources can be assigned to the company officer sector boss who is already in that position. Building the local organization from the onset of the local incident (all incidents ar local) eliminates the need to play organizational development "catch up" later on (and keeps the local incident a local incident). In expanding situations that are not quickly controlled, the IC should attempt to assign subsequent-arriving command officers to transfer command from

Command Safety

Build an Incident Organization

Safety Effect:

the company officer who assumed initial-sector supervision. Command officer sector bosses (geographic) must work in teams of two (chief + aide), and must dress just like the troops, in full gear (PPE), so they can physically operate in the same hazard-zone places as those troops that are under their command. Such a transfer allows the company officer (who was the initial-sector officer) to then concentrate on just supervising their crew, and creates a new (many times command officer) sector officer, who can now focus on just sector officer responsibility. Having the first company officer do double duty is a smart, quick way to begin organizing operations, but the IC must realize how important the need is to create full-time sector bosses. Forecasting as early as possible the number of sectors that are going to be needed will help identify how many company/sector resources will eventually be required. Having the IC quickly do the sector forecast is a critical safety factor. A major problem occurs when there is not an adequate number of command-level officers available, particularly during the hazardous beginning period of operations, to effectively develop an adequate incident organization, that can safely protect the firefighters. The days of dispatching one lonesome response chief in an old beat up sedan must be replaced by the consistent initial response of an aggressive team of trained, experienced command players, who focus on creating a well-practiced, team-oriented command

Command Safety

Build an Incident Organization

Safety Effect:

organization, that is quick enough and big enough to manage both the fire fight and the operational survival needs of the troops. Having an adequate number of command officers on the scene and assignable becomes a huge operational and safety factor. The IC uses those officers as they are needed. Just because they are present doesn't mean they always must be managing something (or somebody). They should stage, stand by, and shut up (we mean that in a nice way) until the IC assigns them. Once a command team is in place and operating effectively, it's okay for officers who outrank the command team members to quietly monitor the command operation. Having the big bosses close and quiet should increase the command team's confidence, and the overall strength of the entire operation. There is no stronger compliment for the senior guys than watching a capable group of their officers (who they trained) effectively work the command system, and after it's over, to mess up their hair and tell 'em, "You done splendid." Having a tactical reserve of all kinds of resources (including bosses) is a smart way to do business... the opposite (too little/too late) is also true. As the incident progresses, the IC will need regular reports (condition/progress/exception/completion) to keep the plan current. Many of these reports come from sector officers. The IC needs to build a big enough organization to find out all of the "unknown" critical information, based on an effective awareness and understanding of the inventory of the critical factors that go with that type of incident. This inventory is the result

Command Safety

Build an Incident Organization

Safety Effect:

of having a fast, practical, critical factor management system in place, along with experienced command team members, who have developed a standard set of incident inventory "files" in their noodle that go with that type of incident. As we assign and use sectors over time, we develop a regular way to divide (sectorize) incidents. This creates dependability and predictability. As an example, if we are conducting offensive operations at a multi-floor building, we will assign sectors to floors--if the situation is defensive, we will assign sides. While we should always maintain a certain amount of flexibility, it's a good thing if everybody operating at an incident can accurately predict, ahead of time, basically how the incident will be divided up and managed.

Command Safety

Build an Incident Organization

IC Checklist:

☐ Accomplish effective delegation and span-of-control management through early sectoring.

Safety Effect:

The IC can become quickly overwhelmed with trying to manage all the different pieces and parts of a particularly fast-moving, expanding incident. Developing an incident organization, and delegating operational responsibility to sector officers, is a major tool the IC uses to prevent becoming overloaded, and to continually support (and protect) their strategic-level capability. This delegation is another way the IC maintains overall control of the incident. There can be a huge amount of operational detail that must be evaluated, decided upon, ordered, and then managed all over the incident site. The IC will be quickly (and completely) overwhelmed if they attempt to personally manage that overall level and amount of detail. Sector assignments become the IC's detail delegation life preserver--simply, managing the local details, that are present in every sector assignment, automatically become a regular, expected part of that sector assignment. Huge management and safety problems

Command Safety

Build an Incident Organization

Safety Effect:

occur when the IC attempts to micromanage the tactical/task-level details, and such over control (i.e., concentrating on task-level details) will quickly overwhelm and eliminate the IC's strategic-level capability. The IC needs to quickly delegate the direct management of the tactical pieces of the fire fight to the people who are in a better position to handle them--sector officers. These decentralized management partners allow the IC to set up, operate, and stay on the strategic level (and to avoid drowning in detail "quicksand"). This improves the safety profile of the event for everyone. Having sectors in place gives the IC more time to spend monitoring tactical radio channels for critical communications, evaluating the effectiveness and safety of the attack, and making any needed strategic adjustments. Sometimes companies who are assigned to a sector officer will go around their sector boss and communicate directly with the IC. This can happen when the sector just got set up, and all the resources in that sector have been commoing with the IC, and are not aware they now have a new sector boss. Other times, insecure company officers who never had a dog in their youth want to commo with the grand exalted big boss (IC). There isn't much use in creating the incident organization, if we don't use it, so when this happens (for whatever reason), the IC needs to redirect (nicely) that company to their assigned sector officer. Having a manageable number of task-level responders directly (and continuously) connected to their assigned sector boss becomes a major safety

Command Safety

Build an Incident Organization

Safety Effect:

capability, particularly when things become unsafe, and we must react quickly. The IC must be the organization sheriff who continually evaluates if everyone is effectively oriented and safely connected to the incident management formation (i.e., organization) that is in place. While the basic sector approach is pretty simple (why it works), it is a major part of the incident safety system. The IC uses sectors as regular organizational components to protect the hazard-zone workers. Geographic sectors provide direct operational and safety supervision. Safety, accountability, rapid intervention, rehabilitation, access control, and resource sectors provide essential welfare-related functional (as opposed to geographic) services to the operational part of the organization. The IC must set these sectors up in a timely and effective way, to cover the operational and safety needs of the incident.

This sectorization also creates the basic capability for the IC to always stay inside the CP throughout the incident. Having an effective, incident-wide "overlay" of assigned sectors, in place and operating, becomes the tactical (and management) layer of the organization that effectively can connect the task level (fire companies) below them, with the strategic level (IC/command team) above them. Having all three levels in place, and effectively operating together, is a thing of beauty and a joy forever.

Command Safety

Build an Incident Organization

IC Checklist:

☐ Correctly name sectors and landmarks.

Safety Effect:

Safely conducting fire fighting operations involves our responders taking a trip (entry) to their assigned work location, doing their job, and (hopefully) taking a return trip back out (exit) of the hazard zone. Like any other trip, it's lots easier if we know where we are going and if we can identify the significant people, places, and things along the way. Our excursion into and out of the work site (hazard zone) is a lot different from most trips because of how fast we travel, and what can happen to us if we get stuck or lost. In order for us to travel safely, it is important that we identify the basic route to our destination, and that we attach regular names to standard parts of the fireground, so that we can assign bosses (sectors) and workers (companies) to those places. The IC must always be aware of the location/safety dynamics of the "attack highway"... the starting point on the outside of a structural fire, the place where the work will occur (destination), and the ongoing integrity of the entry/exit path between those two points. Maintaining this pathway must be a major strategic consideration. Lots of unsafe and painful stuff happens to us when we become confused about where we are, in relation to the basic incident geography, particularly when that

Safety Effect:

confusion affects the IC's ability to control the position and function of the troops operating in the hazard zone. Naming the sides of the building A-B-C-D, floor sectors by floor number, identifying building openings (mostly doors and windows), hallways, stairways, driveways, parking lots, multiple vehicles, and any other significant building/area/access feature or characteristic becomes critical to our safety. Sometimes we should physically mark (chalk, crayon, tape, rope, spray paint, magnetic signs, etc.) those significant parts of the operational area to avoid confusion.

Using these standard location designations becomes the basis of the IC drawing a simple, basic TWS map (sketch) of the area, and must become a major part of developing the incident organization. The IC should create both geographic and functional sector assignments based on how the incident area is laid out. These assignments should connect to an inventory of the significant places in and around the incident, and based on conditions, what needs to be done in those areas. Making the connection between areas and assignment becomes the basis of effective command and control. Using standard sector names that connect to that inventory creates an understandable assignment to a sector officer and the companies who are told to report to and operate in that sector. This standardization (of location/assignment) becomes a major part of how the IC protects those workers. Back-up lines, safety officers, and

Command Safety

Build an Incident Organization

Safety Effect:

RIC teams become regular tools the IC assigns to strengthen both the safety of hazard-zone operating positions, and to protect and retain the safety of the routes in and out of the hazard zone. The IC must not get sucker punched by what appears to be a simple area layout, in the beginning of the event. Not dealing with a lot of basic stuff in the front end can quickly become very complicated later on, and is what typically screws us up. Operating in a five-story building, where the inside guys think they are two floors from the top, when they are actually two floors from the bottom can absolutely wreck their day when they are in trouble and the rapid-intervention cavalry can't find them, based on where they say they are.

Note:

We should not be distracted by how simple it sounds to effectively name the significant incident features. In almost every (fast and dirty) operational event, there is some level of confusion about how we identify these people, places, and things. This confusion becomes a part of almost every incident critique. The difficulty in getting the names attached to important places, and then getting everyone on the incident to effectively connect (and call) everything the same, is an ongoing, big-deal challenge. Sometimes, the difficulty is just an inconvenience that creates a survivable level of confusion. Other times, the difficulty can be fatal. We must make a bigger investment in developing our geographic/functional incident "naming" system before, during, and after the event.

Command Safety

Build an Incident Organization

IC Checklist:

☐ Assign and brief sector officers.

Safety Effect:

The IC builds the incident organization by deciding on the places and functions that must be covered to safely handle the situation, and then assigning sector officers to these areas/activities. Training every company and command officer to serve as a boss of all the standard sectors creates the capability for the IC to develop an incident organization, as the (hopefully interchangeable) responders arrive on the scene in whatever order. Initial-sector assignments cause the IAP to actually go into action and create the on-scene site bosses, who manage the manual labor and related safety support that solves the incident problem. The IC must quickly include the location/task/objective in assigning the sectors, to create an effective beginning to the work of that sector. Once the sector officer is in place, the IC must then critically listen, evaluate, and process information coming back from the sectors. This information becomes the basis of keeping the IAP current and safe. Being able to assign the first company officer in a tactical position, as the initial sector officer, and then later (if the incident problem continues) to upgrade that position to a command officer, creates the capability to quickly establish a sector, and

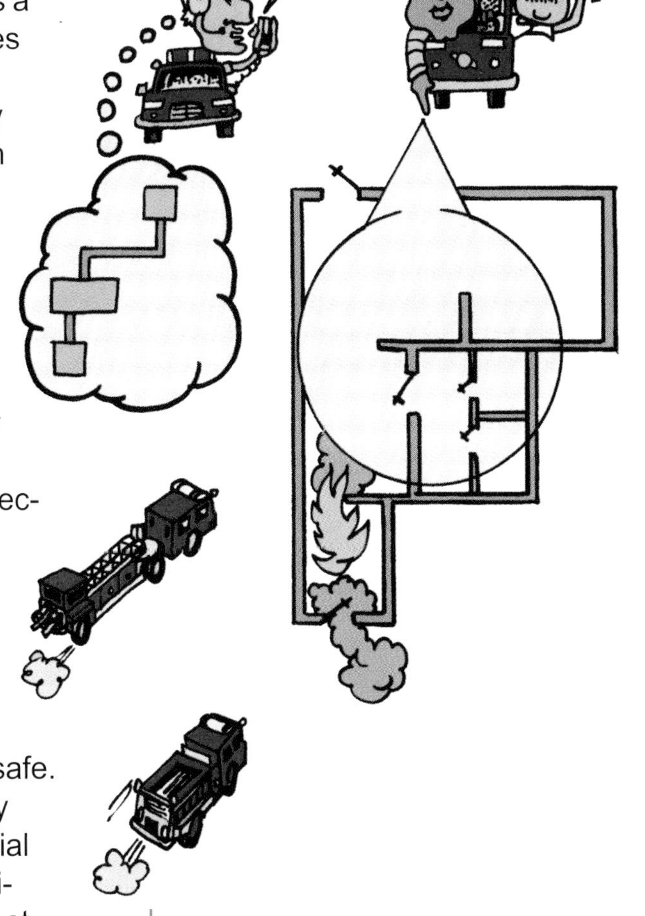

Command Safety

Build an Incident Organization

Safety Effect:

then to later back up that company officer. The IC must realize that when a company officer is working as the initial-sector officer in a location/function, they are generally doing double duty--i.e., that officer is simultaneously doing some basic sector functions, and looking after (in some way) their company. When enough command staff arrives and can be assigned to relieve the initial-company officer sector boss, the level of supervision increases. Now the company officer can go back to managing just their company, and the command officer sector officer can now concentrate on doing just the sector functions. Using the company officer as a sector boss to begin with is still the strongest (and quickest) way the IC can get an increased level of initial supervision, information management, and safety awareness all over the incident site. Building on this company-officer beginning, by upgrading to command officers, works well because there is sector supervisor in place when the command officer arrives (in the sector), and there is some sector-level supervision in place to transfer (it's difficult to transfer something that hasn't been established). Our experience consistently reflects that while hard-working company officers do a good job as initial-sector bosses, when we are able to assign command officers in critical sectors, we create the capability to shift to a more tactically-oriented view, rather than a task-level view... this is a big deal in the effectiveness of our overall organizational approach. As incidents escalate,

Command Safety

Build an Incident Organization

Safety Effect:

the IC must continually reinforce strong, two-way communications by using alternate radio frequencies for functional sectors, develop a CP staff to assist with ongoing communications, and be continuously available to commo on the primary tactical channel with the hazard-zone sectors/companies. Crowded, overloaded, completely screwed up radio frequencies always seem to be present when firefighters get beaten up or killed. These are the most awful situations an IC can face. Managing, directing, and controlling radio traffic is a major way the IC prevents the nightmare from happening. Anyone who thinks that someone saying, "We have a firefighter missing" on the tactical channel, and then having everyone just automatically clear the air for the IC and the RIC to operate, lives in a flat-ass dream world. These are (typically) massively fouled-up situations that must be prevented, because they create so much confusion, trauma, and stress that they probably can never really be effectively managed, particularly the longer they go on. This is why the IC must manage in a way that always prevents them from happening.

Command Safety

Build an Incident Organization

IC Checklist:

☐ Limit units assigned to sectors to five (5).

Safety Effect:

The IC must avoid assigning more resource to a sector than the sector officer can effectively manage. When this happens, it amounts to drowning in resources. A major part of the sector officer's responsibility is to manage the safety of all assigned personnel. This is impossible, if there are too many resources assigned to that sector officer. The smart money sez that trying to manage more than five (5) companies starts to stretch the span of control of particularly very busy interior-attack sectors, who are typically managing under active, fast, and dirty conditions. Creating standard teams (like two engines, one truck, and one BC), who routinely work together, is a familiar and effective organizational/operational plan, number, and configuration. When there is more work to be done, or more area to cover, and the sector already has five companies, the IC should create another sector. The same approach to limiting five units to one boss also applies above the sector level. In big deals where the IC gets five sectors going, they should start to think about creating branches to directly manage

Build an Incident Organization

Safety Effect:

the sectors, and to maintain the IC's effective span of control. Basic incident organizational design (and philosophy) is like managing kittens in a basket. When there are too many of them, it is impossible to keep track of them, and to at least keep them in sight of the basket. As they add kittens, they must add more cat herders.

Command Safety

Build an Incident Organization

IC Checklist:

☐ Serve as resource allocator to sectors.

Safety Effect:

Once the sector officer is assigned and briefed about what to do in their sector, the IC then supports the sector by serving as a coordinator and as a resource allocator. If the sector officer needs more resources, they go through the IC to get them. The sector officer manages the work activities in the sector and provides progress reports to the IC. This puts the IC in a strong position to react to reports and manage the incident on the strategic level. Using the IC as the resource "gate" creates the capability for the IC to maintain an overall awareness of the ongoing "who is where, doing what, assigned to what boss, when did I talk to them last, are they okay?" status of the incident. Assigning arriving responders (from their staged position) to their initial place in the incident organization is a way the IC builds and maintains an inventory accounting and tracking system, from the very beginning of the incident, and then continues that system throughout the event. Once resources are allocated, they must be managed. The IC must guard against assigning responders to go to work, and then not continue to manage those work assignments. Telling a company to go to the inside and do search and rescue (as an

Command Safety

Build an Incident Organization

Safety Effect:

example) is pretty easy--the tough part is tracking their general location, maintaining an awareness of their progress and welfare, and moving them if conditions change for the worse. The standard work cycle is a major way the command system "captures" an operating unit, and then accounts for that unit all the time they are on the scene. The initial assignment provides the front end for that cycle. Maintaining control of that standard work cycle is a major way the IC eliminates free-lancing and is continually aware of the location, function, and welfare of the troops.

Command Safety

Build an Incident Organization

IC Checklist:

☐ Build a command team.

Safety Effect:

Active, fast-moving incidents that occur in tactically significant buildings/areas quickly create a very difficult (to say the least) command and control challenge for the IC. Major safety problems occur when the IC is not able to maintain command and control of the position and function of responders, in and around the hazard zone. These difficult (very local) situations are currently underway and can many times require a fast, large deployment of lots of resources. What "very local" means is that a hazard zone is present, and that hometown resources must be quickly deployed to make the out-of-control conditions (that make it a hazard zone) go away. The problem(s) is not abstract or academic--anyone who requires rescue needs real-live responders to quickly go inside, and either protect them in place, or take them outside; anything that is on fire (structure/interior finish/contents) will stop being on fire, only if those same real-live firefighters hold the fire's head under water. A big part of both the safety and operational challenge is that all this up-close and personal hometown action must be done quickly.

Command Safety

Build an Incident Organization

Safety Effect:

The size and scale of these local incidents can quickly exceed the capability of a single person (IC) to manage by themselves. These events require a local level of reinforced command that must be implemented almost instantly by the on-duty homeys. It's only interesting to deliver a big-time command team ("overhead") that shows up twenty minutes after the last combustible is vaporized... for us to wait for the big dogs to arrive is to surrender the incident to the hazard. The hometown IMS must be able to quickly assemble a command team to support the IC. How that local team routinely comes together and operates must, very simply, occur (if it is effective) before the building burns down. The idea of front-end loading a large amount of predetermined, practiced command team effort into the incident response must occur in the earliest period of our operation. The ability to capture control of the front end of the incident is when we stand the best chance to overwhelm a structural fire, and it is also the period where we have the best chance to injure and kill the hazard-zone workers. The windows of opportunity (and effectiveness) get smaller (and go away) pretty quickly. When we put the fire out, everything gets better... and safer. Local hometown structural firefighters must be packaged up, and do some fast-and-dirty heavy lifting--forget big-time wildland command configurations (where they actually fly the overhead team in from Minnesota) two days after the fire starts. These are the people who are the very best at managing large-scale, long-term, big deals, and there is a lot that we locals can (and do) learn from their system and how they operate.

Command Safety

Build an Incident Organization

Safety Effect:

But for us homeys, the biggest offensive structural fire fighting challenge for us is to quickly get two to three attack teams in the right place, to quickly kick the local fire's ass. The command team assists the IC with anything (or anybody) that gets in the way of the IC's ability to manage tactical, task-level units operating in the hazard zone. The command team is designed to "front-end load" a quick, reinforced, street-oriented, team-based response that occurs, as early as possible, in the incident. Firefighter safety is a major focus of the command team. Setting up sections (discussed later in this chapter) works well on extended operations, but typically takes too long for fires that are wrecking (right now) the inside of a single building, and starting to extend to exposures. Huge safety problems occur when firefighters are operating in the hazard zone and the IC is overwhelmed--particularly in the front-end of the incident. The major focus of the command team is to create, as quickly as possible, a big-time command concentration on controlling and protecting the position, function, and welfare of the troops in the hazard zone, who are trying to put a quick hit on the fire.

The standard IMS sections (operations/planning/logistics/administration/safety) are designed to create an expanded organizational capability on large, campaign (long) incidents. They work very well on these kinds

Command Safety

Build an Incident Organization

Safety Effect:

of situations. They are designed to decentralize the command organization by delegating in a standard configuration, the major management activities that are required to effectively conduct big-deal command and operations. Sections were originally designed (and applied) to manage large wildland fires that cover a big area, require a lot (a lot!) of responders, that typically burn a long time, and have twenty-four hour incident action planning periods. Through the years (of IMS development), our service has applied these same sections to a full range of types of incidents ("all risk"), and they are an extremely effective management system when applied to large, extended events. The command team approach is distinctly different (actually the opposite) from the use of the four standard sections. The command team is set up in the very beginning of local incidents and is designed and used to concentrate and centralize (not decentralize) the command function. The team is assembled and staffed as quickly as possible, by local street responders, and attempts to provide a hard-hitting level of command in the earliest period of the incident, where (and when) local responders have the best chance of pulling off the basic tactical priorities. If the efforts of the command team do not effectively stabilize the incident problem, then the command team "evolves themselves" into the standard sections, and uses them as the organizational basis to conduct campaign operations.

Command Safety

Build an Incident Organization

Safety Effect:

The authors have observed the "we do that same thing--we just call them sections" response from those who are familiar and comfortable with sections, during the discussion about command teams. While this is an understandable reaction, it is not an accurate one--with local-level command teams, the support arrows all point in to the IC; with sections, the arrows point away. This is not meant as any criticism of sections. While there is a place in an evolving, large-scale, long-term event where the command team transitions into sections, the two organizational forms are designed and used to do a different but very connected set of management things, at different stages of the incident.

A major IMS safety objective is to continually create the correct level of command to match the needs of the incident. Having a command level that is too small, or too big, exposes firefighters to an unsafe level of risk. Where there is too little command in place, firefighters are not protected by an effective level of direction, control, and supervision. When this happens, there are more responders operating in the hazard zone than the IC can manage and maintain control over. Conversely, overbuilding the command system in the beginning of the incident, typically takes a length of time to assemble and put in place. That causes hazard-zone workers to be unprotected while the big-time organization is being put in place, during the critical (and very dangerous) stages of initial (particularly offensive) structural fire fighting operations. These local incidents generally require responses that range from first-alarm levels of

Command Safety

Build an Incident Organization

Safety Effect:

five to six companies, to second and third alarms up to about twenty companies. To manage those response levels, a quickly assembled command team (IC/SO/SA) made up of local street responders (mostly command officers), who are assigned as an integral part of the dispatch(es), become the quickest, most effective command level designed (and customized) to fit that local, multiple-alarm response level. For these incidents, it is slow and cumbersome to attempt to either increase the size of a system that is inherently too small, or to somehow miniaturize a system that is designed to manage disaster-size incidents over an extended period of time. The critical safety period for local fire fighting operations begins in the very beginning stages of operations. It is difficult and dangerous for locals to try to apply a non-local level management system to an event that exposes our workers to a huge risk level. For a room-and-contents fire in a single family, a lone IC works well. If we have a 200-acre expanding wildland fire in heavy fuel, send for and set up the command army--for up to twenty companies operating inside the burning Smith Hardware Store down on Main Street, quickly set up a local command team that will effectively manage and protect the fast-moving attackers, who recently arrived on the low

Command Safety

Build an Incident Organization

Safety Effect:

numbered engines and trucks (like E-1, L-3, etc.). That same command team is also the best shot that we locals have to effectively evolve into the standard sections, when they (the command team) identify that the incident will require an extended effort.

The command team assembles and operates together inside the CP, as quickly as possible, and is made up of the following three standard positions:

1. IC:

- serves as IC--personally and directly performs the eight standard command functions... performing these functions becomes the basis for the IC completing the strategic operational and safety function for the entire incident, and is the major focus of this entire essay
- calls themselves and answers to "Command"
- maintains primary, two-way contact with haz-zone companies/sectors on the primary tactical radio fire frequency
- talks to/listens to the commo center
- monitors and manages the overall strategy and IAP.

Command Safety

Build an Incident Organization

Safety Effect:

2. SO:

- sits right next to the IC and serves as the IC's " intimate advisor"
- protects IC from distractions (physically, if necessary)
- directly supports the IC
- maintains the TWS
- coordinates requests for resources between the IC/ staging
- continually challenges the currency and safety of the overall strategy and the IAP
- evaluates the immediate and future safety of the haz-zone workers
- manages the IC CP (office)
- keeps others from face to face with the IC (i.e., buffer/barrier/ blocker)
- evaluates how the existence, intensity, and changes in the critical factors are influencing the strategy

Command Safety

Build an Incident Organization

Safety Effect:

- determines what the critical factor trends are and what will cause us to change the strategy.

3. SA:

- looks at the "big picture"
- confirms or changes strategy/ IAP

- expands the incident organization as needed
- coordinates between IC/ SO and section positions
- continually reviews and "plugs up" holes in the adequacy of the resources that execute and support the IAP
- coordinates with and assists other agencies
- informs and deals with bosses (administrative and political)
- evaluates effect on community (short/medium/ long term)
- asks: "How will this look in tomorrow's newspaper?"
- connects the command team (and the overall incident) to major community elements.

Command Safety

Build an Incident Organization

IC Checklist:

☐ Build outside agency liaison into the organization.

Safety Effect:

Significant local incidents quickly involve the participation of other agencies within the community. These agencies come to our events to perform their specialties, just like we go to their incidents to assist them by performing what we do best. Agencies like law enforcement, medical, ambo, special operations, electric, gas, water, public works, cable, and telephone utilities routinely respond and operate together, to get all the pieces and parts of the incident covered. We should look out for them, and they for us, while we are on the scene together. The IC should automatically build in an agency liaison into the incident plan and organization. Some incidents require the major agencies to shift from being in command to supporting another agency that assumes command, as the event evolves (example, aviation incidents starts with Fire, then EMS, then crime scene, then National Transportation Safety Board (NTSB), then Aviation Department, then airline, then lawyers). Making this "lead-follow/follow-lead" transition requires strong unified command (positive relationships, developed ahead of the incident), and good playground behaviors (i.e., playing well with others). A major safety factor involves every agency taking the lead expert role in their specialty, and protecting everyone else from that hazard--i.e., fire--fire/

Command Safety

Build an Incident Organization

Safety Effect:

cops--violence/electric--shock/gas--kaboom/hazmat, etc. The IC must lead this process by being the head listener, supporter, and coordinator. Developing the interagency capability for a bunch of aggressive responders from different agencies to come together out on Main Street and play nicely, help each other, and to use their specialties to mutually solve a fast-moving difficult local situation is a challenging, multiple-agency task. The IC must do standard liaison first, then check back as required, to be certain each agency is informed, integrated, and okay. Many of the problems associated with this challenge directly relate to the safety and welfare of all the players, who show up at the game. A continual cluster-like outcome will occur until and unless a strong IMS is in place, supported and led by smart, tough bosses who both create the process and model positive behaviors.

Command Safety

Build an Incident Organization

IC Checklist:

- ☐ Operate on the strategic level--support tactical and task levels.

Safety Effect:

The IC must perform with the personal discipline to operate on the strategic level. Sometimes, this isn't easy. Actual incident conditions/operations create a variety of distractions that can pull the IC away from playing their command position. Leaving the CP, wandering around "evaluating" stuff, and giving random orders, micromanaging excessive detail, showing younger, lower-ranking responders "how to do it," not maintaining a "listening post," and not doing strategic-level documentation all set up big-time command level safety problems, when conditions worsen or tactical surprises occur. The IMS requires responders who are trained to effectively operate on their assigned level and an IC who empowers and trusts them to do just that... operate (i.e., do their job). When the tactical/task level get in a bind and need command to quickly support them, it's pretty tough to get that help, if the IC is 250 feet away from the CP teaching residual hydrant pressure dynamics to a young pump operator. The IC does their command and safety management job by doing the standard functions of command (which are in and of themselves a challenging, full-time job). Tactical and task levels must be able to depend on the IC to be outside (but close to) the hazard zone, inside a CP doing the

Command Safety

Build an Incident Organization

Safety Effect:

command stuff (functions) that protects the troops, who are inside the hazard zone. We have now added a rapid intervention component to our standard, tactical safety approach--these are standby teams that stay outside, close to the action, always ready for rescue. RIC teams are task-level workers, who are physically ready to help inside crews if they get stuck. The IC is the outside person who operates on the strategic level. We should think of the IC, along with the SO and SA as the RIC team on the command level, who together in a highly-integrated way, receives, processes, and manages the ongoing information from insiders, about the safety and welfare status in their area. Sometimes changes in conditions require activation of a tactical-level RIC response into the hazard zone. The IC is the only person who is consistently set up on the strategic level (in the CP or anywhere else) to effectively activate the RIC team, and to provide the overall support that response requires. This strategic command capability must be established and maintained routinely at every event, so it is in place when it is needed.

Command Safety

Build an Incident Organization

IC Checklist:

☐ Evaluate progress reports, assist, and coordinate sector activities.

Safety Effect:

The IC will typically be giving orders and assignments during the front end of the event to get the operation started. Once the organization is in place, the IC should then balance listening with talking (really not a bad idea all the time--everywhere). Geographic and functional company/sector officers will be making progress reports about work efforts and conditions in their sectors. The command system puts the IC in a position to make operational changes based on the changing conditions. A major command function involves the IC maintaining the capability to put an incident organization in place, and then to use the CP capability to receive (listen!), evaluate, process, and then to quickly respond to condition/progress/exception/completion reports. The IC must receive and process information from all the critical places, and develop an ongoing evaluation to determine the overall status of the incident. If conditions get better, the IC presses forward. If conditions get worse, the IC needs to reevaluate that part of the attack. Based on how they are managed, the activities of different sectors may help or interfere with each other's operation and objectives. How sectors interact and connect with each other becomes a major coordination activity for the IC. The IC must vigorously coordinate

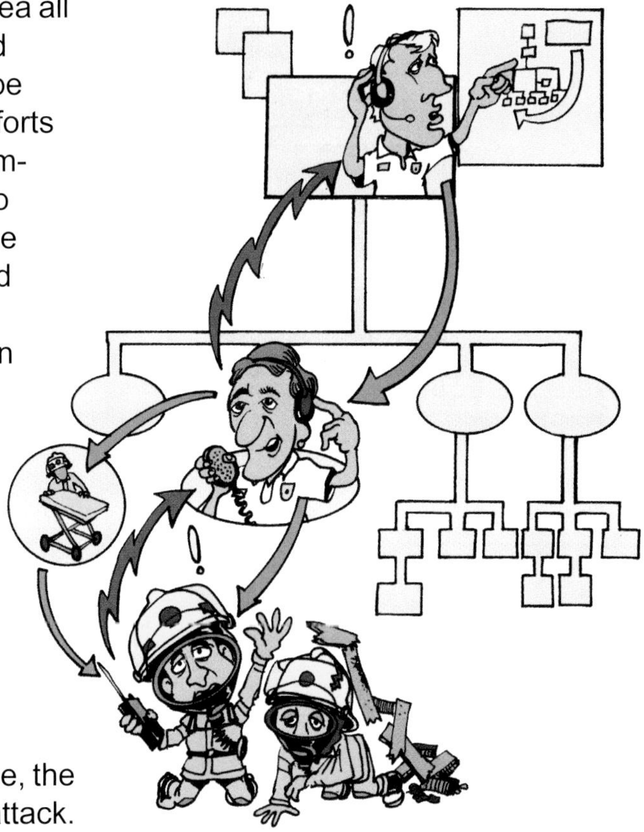

Command Safety

Build an Incident Organization

Safety Effect:

and manage the incident organization at situations, where sectors are working at cross purposes. Many times, these inter-sector difficulties cause big-time safety problems. Having outsiders blast insiders, crossing inside hose streams, having insiders get stuck because of unventilated operational areas are examples of situations, that require strong command support from the IC. Sometimes, this means the IC must order a participant to "stop doing what you're doing and start doing something else that is safer and more effective."

Command Safety

Build an Incident Organization

IC Checklist:

- ☐ Implement management sections and branches to support operational and command escalation.

Safety Effect:

All incidents are not created equal, and how we organize the strategic level must match how each incident was, in fact, created. It becomes a big-time responder safety and operational effectiveness issue how the match between the incident profile and the incident organization is achieved. As we have stated throughout this essay, incidents involving five to six companies require a command system that matches that level. Those with fifteen to twenty companies need their own command level, and big-deal campaign events, requiring twenty-five plus companies, begin to look like long-term, local disaster level responses, that begin to require an Emergency Operations Center (EOC) to coordinate with the outside world. For most of us, these events would move through being local (five to six), mutual aid with neighbors (fifteen to twenty), and regional (twenty-five plus). A major IMS capability is that the system can expand and contract to fit the situation it is applied to. A major IMS capability involves using sections and branches as standard organizational components, that are used to expand (significantly) the "upper-level" command part of the incident organization. Using these components creates the capability to capture

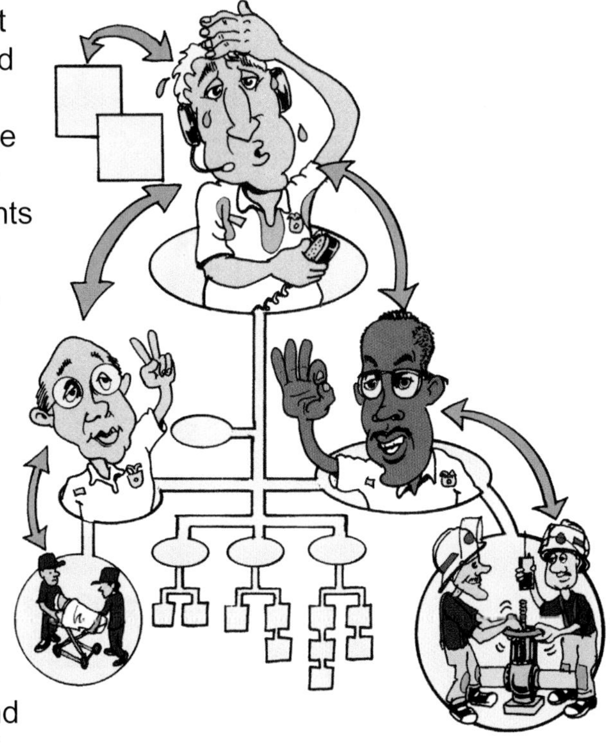

Command Safety

Build an Incident Organization

Safety Effect:

and maintain control of large-scale situations, that will go on long enough to set up these "bigger" IMS parts. Matching the command system to the level, area, and severity of the situation becomes a critical safety factor... any time/place the command system is behind the organizational power curve (size, duration, complexity) creates a problem in protecting our fire-fighters. Branches are established to manage multiple geographic sectors at incidents with big areas and lots of sectors. Branches are also used to manage a major function, i.e., fire, medical, haz mat, evacuation, etc. Sections (not sectors... sections) are implemented to assign an upper-level manager to the major support/operational areas, in incidents that will require a lot of resource over an extended period of time. Regular section designations are operations, logistics, planning, administration and (a new one) safety. The safety section has reinforced the focus on managing and supporting worker welfare and survival, and creates a section-level CP position to focus on and manage the on-scene safety officers, accountability, RIC, structural evaluation and an ongoing review of the safety and survivability of the IAP. In situations where a command team (IC/SO/SA) has been established, and the incident continues to escalate, the command team must evolve themselves into the sections that will be required to conduct longer-term management of a continuing event. The three-person command team is quickly assembled to front-end

Command Safety

Build an Incident Organization

Safety Effect:

load the operation with fast, strong command--the sections are used for longer-term campaign type incidents. Most of the time, on a local level, a command team is set up first and then evolves into a section-based team, in situations where the fire simply keeps burning (and burning, and burning).

Command Safety

Build an Incident Organization

IC Checklist:

☐ Use organization chart as communications flow plan.

Safety Effect:

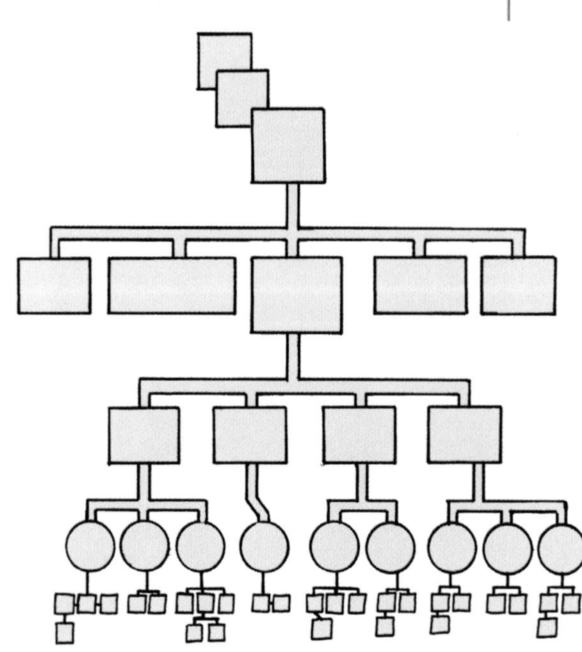

Throughout this essay we have worn out the reader with a couple of recurring themes. A major one involves us being able to decentralize how we locate, organize, and balance ourselves on the strategic, tactical, and task levels throughout the fireground, so the IC (centralized and STABLE... as opposed to mobile) can receive reports from all the critical places and functions (decentralized and mostly mobile). A major objective of this approach is to create an effective two-way information flow. While this approach creates a positive capability, it can also create its own set of problems. In recent times, we have provided portable radios to lots of responders... everyone operating in the hazard zone should ideally have their own personal portable radio. This creates the capability for every hazard zoner to commo with the command team about ongoing incident action, and if they have any critical safety needs. Using the standard radio signal "Mayday" creates the highest and most urgent commo status (and response). A possible downside of individual radios is that more radios potentially can equal more nonfunctional talking. If there are (as an example) three companies operating in a particular tactical position, all three of them may have the tendency to communicate (sometimes chat and jabber) over the radio. When the IC assigns sector

Build an Incident Organization

Safety Effect :

responsibility to that area, the multiple jabberers are then assigned to one sector officer who deals with their companies face to face, and becomes the local "radio agent" with the IC for that position/function. Multiply this by the number of tactical positions and assigned resources to those areas, and we can see the overall very positive effect that sectoring has on communications (and safety). This approach becomes even stronger and more effective when the IC uses the organization chart as the template to structure incident communications. This requires the IC to communicate with sectors using radios, and sectors and their companies doing mostly face-to-face contact/commo for routine information exchange. Everyone uses a combination of mobile and portable radios as a standard way to transmit, retransmit, and coordinate "emergency traffic," whenever it is necessary to report critical safety conditions, and to request and manage the support required to deal with those situations. Many safety problems "show up" first as some sort of communications difficulty--wherever we say commo, we should also say (or think) safety. When commo gets screwed up, the IC had better straighten it out, because safety is next.

Command Safety

Build an Incident Organization

IC Checklist:

☐ Allow yourself to be supported in the process.

Safety Effect:

A major IMS approach (overall) involves starting initial operations in a variety of ways (and places), to quickly get the problem solving going, and then continuing the response to reinforce and support those efforts. Starting operations and command together (early) creates an effective, under control, local reaction that many times quickly solves the incident problem. This same routine should apply to how command is initiated in the beginning and then expanded, as the incident response escalates. Major safety problems occur when the level of overall command capability falls below the management needs of the companies and sectors working in the hazard zone. The deployment system must provide for the response of multiple command officers, as the incident escalates. It is a very effective (and safe) practice that response chiefs have full-time assistants who drive, assist with command, and support communications. A chief and aide can be assigned to manage a sector. Having a two-person team arrive together qualifies them to go into the hazard zone (no onesies in the hazard zone). A lot of response chief officers are not provided full-time aides. Simply, this is a current reality in our many times under-resourced business. Systems without aides should develop SOPs that provide for on-scene responders to join

Command Safety

Build an Incident Organization

Safety Effect:

the IC at the scene, and to serve as operational helpers. Other response bosses can join each other, or other special responders can serve as aides. Response chiefs who respond by themselves must be very careful driving to the scene, and must concentrate on getting through traffic, and not be distracted by radio traffic from the scene. This is one of the advantages of having a command assistant who can drive, while the chief listens to radio commo, starts a TWS, accesses preplans, manipulates mobile data and computer terminals, and generally begins to develop an awareness and focus on how the incident is shaping up. The command system is generally established in an early-arriving response vehicle--like an engine or a ladder truck by an initial-arriving company officer. When a command officer arrives, command can be moved to a quieter, smaller sedan or SUV, that can be moved back a bit to get out of the way of operational units (engines and ladders), and to achieve a better vantage-point position. Specially designed CP vehicles create another level of position upgrade--these rigs are peachy because they have room for more staff, lots of command tools (toys), and their own supply of Snickers bars. While a lot of fancy rolling stock and hardware is nice, it's not necessary. The point is that the IC must assemble a command team in as good an environment, as quickly as possible, to manage the operational and safety affairs of the firefighters. Building a sector system that covers operations, support, and safety becomes the command

Command Safety

Build an Incident Organization

Safety Effect:

partners who cover the whole event. Setting up sectors to cover safety, rehab, accountability, access control, rapid intervention, resources, and structural evaluation become the standard part of the incident organization that focuses on safety and survival.

... Remember, even the Lone Ranger showed up with an aide, who was armed and had a big spotted horse.

Command Safety

Build an Incident Organization

This happens when Organization is done:

- ☐ The IC initiates/maintains a strategic level of command.
- ☐ Geographic and functional sectors are established to cover all the incident operational and support needs on the tactical level.
- ☐ Fire companies are effectively and safely supported on the task level.
- ☐ An effective span of control is maintained on every level by creating an organization to match the size of the incident and the size of the response.
- ☐ The organization is continually adjusted (+/-) to match the management needs of the incident.
- ☐ Everyone has a boss, knows who their boss is, can communicate/ connect with their boss--commo is controlled and organized.
- ☐ The IC can use the incident organization to manage/change the IAP to protect the troops.
- ☐ Effective organization = better control/better control = safer operations.
- ☐ An effective organization is in place to forecast and prevent safety problems before they occur.
- ☐ Effective organization = big-time mess prevention.
- ☐ Safety system predicts/prevents/fixes problems before they occur.
- ☐ Effective action is evaluated, noted, passed on/problems are corrected.
- ☐ Ongoing support is provided for correctly placed players--little room for free-lancers.
- ☐ Tactical safety and operational objectives are completed/customer's needs are met.

Command Safety

Build an Incident Organization

This happens when Organization is not done:

- ☐ The IC does not establish an initial incident organization.
- ☐ Everyone works directly for IC (... dog pile).
- ☐ There is no effective delegation.
- ☐ The IC gets overloaded quickly (span of control = tilt).
- ☐ No effective organization is in place to manage IAP--no forecast of safety problems.
- ☐ No functional coordination between companies.
- ☐ The IC calls for more resources--still doesn't build organization--resources overwhelm IC.
- ☐ Control, coordination commo (... safety) quickly goes out the window.
- ☐ Big commo mess--everyone blabs--IC is overwhelmed... slowly sinks into "commo quick sand."
- ☐ The troops get lost, disoriented, disconnected from safety system.
- ☐ Huge unsafe, confused, cluster----, big unmanageable (sometimes unsavable) mess.
- ☐ The IC stays tactical (sometimes task) instead of strategic.
- ☐ Firefighters end up in offensive positions under defensive conditions... always beaten up/sometimes don't survive.
- ☐ Customer is forgotten in the confusion.
- ☐ We look like idiots--attorney visits highly stressed fire chief.

Command Safety

Build an Incident Organization

COACHING VERSION

"Quickly create a sector-based organization to delegate both area responsibility and major functional responsibility. Be certain you automatically consider the incident problem in relation to all seven sides, and the sector functions that will be required to support the hazard-zone operation, and to eliminate the surprises that conditions in unseen areas can produce. Organize early and aggressively--develop and then (you) create and push your own operational/ command power curve (GPMs vs BTUs) to quickly bash the incident power curve. Establish only the part of the command system that is required to get the job done. Don't get overloaded with excessive organization because you're too big--don't play catch up because you're too small. Expand the incident organization as you call for and assign more resources. Keep the size of the organization always a little bit bigger than the resource level--don't let that size get behind... get the organization going early because in fast-moving situations, it's almost impossible to catch up. Always remember that sectors become your eyes, ears, nose, and gut all over the incident site. They create the capability for you to be inside the CP and to stay connected to the entire incident. Assign the initial unit in the critical area/function as a sector to begin early organization and control. As quickly as possible, replace company officer sector bosses with command-level officers, so those company officers can go back to leading their troops. As soon as possible, assign a SO and SA to develop command team (partners), before you get into trouble. Stay in the CP, listen critically, aggressively support your sectors and let them manage the details of their location/assignment right from where the action is actually going on--put on your "big ears"--manage the strategy--allocate resources--pay attention--emphasize safety--and remember you are always playing a game with the incident power curve and whoever is faster/bigger wins."

Command Safety

Build an Incident Organization

Command Safety

Review/Revision

Command Safety

Review/Revision

Command Safety

Review/Revision

Function 7

Review/Revision

Major Goal

To complete the steps required to keep the strategy and action plan current.

IC Checklist:

- ☐ Regular command system elements established in the beginning provide the framework for midpoint review/revision:
  - strong standard command
  - sectors
  - SOPs
  - risk management plan
  - strong communications
  - standard strategy/action planning.
- ☐ Carry out command functions (one to eight) in a standard order.
- ☐ Receive, confirm, and evaluate conditions--progress reports.
- ☐ Use standard strategy/action plan review items as the checklist for revision:
  - firefighter safety
  - does strategy match conditions
  - 1/2/3 priority progress
  - correct action
  - location of attack
  - size of attack
  - timing and amount of support
  - adequate back up
  - adequate resources
  - have a plan B.
- ☐ Quickly make transitions based on the safety profile of changing/forecasted conditions:
  - feedback from sectors/companies
  - quick evaluation
  - move the troops
  - regroup--go to plan B.
- ☐ Quickly provide “salvage command” if necessary.

Command Safety

Review/Revision

IC Checklist:

☐ Regular command system elements established in the beginning provide the framework for midpoint review/revision:
- strong standard command
- sectors
- SOPs
- risk management plan
- strong communications
- standard strategy/action planning.

Safety Effect:

Having the IC perform the first six standard command functions in order (with lots of cross over/integration/mushing together between and among the functions to meet the management and safety needs of the incident), from the very beginning of the operation, provides the foundation for whatever midpoint IAP revision is required. A major part of any management system involves creating an ongoing approach that requires a boss (in our business we call that person the IC) to always match and connect the overall strategy and the operational plan to current conditions. This is a big (safety) deal. Midpoint review and revisions must be based on the basic idea that initial operations are the result of a plan that is made up of standard elements. It's pretty tough to evaluate and make an adjustment (in the plan), in the middle of the operation if the plan was never developed (using the standard pieces), and put in place in the beginning of the operation... simply, it's pretty much impossible to revise a "non plan." How the system gets set up in the beginning and how it gets revised (as required) in "the middle"

Command Safety

Review/Revision

Safety Effect:

must be pretty simple (and closely connected), if it is going to work under actual incident conditions. When the system is set up in the very beginning of operations and effectively in place, the IC is in control, sector officers are in their assigned places/functions in the key tactical positions, running operations according to the IC's IAP (based on the current strategy). The IC is linked to the fire companies/ sectors over the tactical radio channel. The basic operational tactical priorities (rescue/fire control/property conservation) get completed within this standard framework. The system creates the basic command components that set the IC up in the best strategic position to find out any critical safety "unknowns" and then quickly and safely react to them. Addressing the "knowns" and assigning units to find out the "unknowns" becomes the major way the IC determines and identifies the risk. Companies are assigned in manageable numbers (i.e., within an effective span of control) to the different sectors. This simplifies and streamlines reporting progress toward the standard incident goals and makes it possible (when it is necessary) to quickly and efficiently move companies out of harm's way. Having the overall strategy and the IAP transmitted to all the different sectors will be the basis for condition/ progress/exception/completion reports along with ongoing condition reports that the IC will use to revise the IAP, if the initial attack does not solve the incident problem.

Command Safety

Review/Revision

Safety Effect:

A major objective of the IC automatically setting up and doing the standard (eight) functions of command is to create the capability to move the troops to safe positions/ conditions (when necessary), to always protect them from changing conditions that can exceed the capability of our basic safety system. The IC must operate at the strategic level and be in a position to use all the pieces and parts of the IMS to directly control the position and function of everyone who is operating in the hazard zone. Those operating everywhere else (i.e., not in the hazard zone) are mostly directed automatically by SOPs. Using SOPs to manage the non-hazard zoners frees up the IC to concentrate on the assignments and activity that is going on in the hazard zone. Local SOPs typically provide for a standard array of support functions to be automatically set up to provide assistance to the hazard zoners. Functions like staging, safety, rehab, accountability, access, and resource control all become the "ground crew" that keeps the hazard zone "flying"--these sectors are set up and directed mostly by standard procedures and responders who are preassigned to those support activities. When they are established and operate automatically, they become a huge help to the IC because they require very little direction (from the IC) to get them in place and operating. The regular, local deployment SOPs are also designed to create a steady stream of responders who are coming into the incident and "quietly" standing by, available, and ready for the IC to assign them where they are needed. Standard dispatch, staging, assignment, and accountability procedures create these regular (under-

Command Safety

Review/Revision

Safety Effect:

standable, predictable, dependable) front-end behaviors for responders coming into the incident. Physical entry into the hazard zone should be regulated by the following standard (short/simple) requirement:

- Everyone who goes into the hazard zone must be:
  - fully protected
  - with their intact crew
  - assigned to a sector (by the IC).

The IC must integrate the regular strategic-level safety routine into the standard command functions to provide for firefighter survival and welfare (which is the major objective of all the detailed stuff described and discussed in this essay).

The reason those who serve as ICs should examine and internalize all this command safety "stuff" is that the major, absolutely most important element in the structural fire fighting game is the lives of our firefighters. If they (the firefighters) must actually and routinely risk their welfare in the hazard zone to protect Mrs. Smith, it is appropriate that the IC should spend whatever time, and invest the energy that is required to learn, apply, and refine the standard command safety routine that is required to protect those hazard-zone workers. If those who are called upon to provide command are not willing to take the time and extend the effort to effectively learn their command and safety role, they should stay home and off the fireground.

Command Safety

Review/Revision

IC Checklist:

- ☐ Carry out command functions (one through eight) in a standard order:

  1. Assumption/Confirmation/Positioning
  2. Situation Evaluation
  3. Communications
  4. Deployment
  5. Strategy/Incident Action Planning
  6. Organization
  7. Review/Revision
  8. Transfer/Continuation/Termination.

Safety Effect:

The standard command functions become the IC's basic job description. Performing the command functions provides for both strategic-level management of the incident response and for worker safety. Integrating the standard, strategic-level safety routine into the regular command functions eliminates the need for the IC to first remember and then to actually do two separate routines, which is difficult under typical incident conditions (fast, confusing, dangerous). Doing the command functions (in order) provides a standard progression of regular management activity that front-end loads command and control into the beginning of the event, and then provides the foundation for whatever midpoint (or later) revision is required (as we just said, "It's hard to revise a non-plan"). Many times, the initial IAP (fast, effective, offensive action) effectively controls many local incident problem(s) and quickly reduces and eventually eliminates the hazard to the workers (and to

Safety Effect:

everybody and everything else). In these frequent (local) quick/offensive cases, there is very little need for plan revision... simply, put out the fire, and everything gets better. In some other situations we extend the initial IAP, but it does not control the incident problem(s). When this happens, the IC must then somehow revise the IAP to match whatever part of the incident that is out of control and is making the workers' lives unsafe. Big problems occur when operational revision is required and the IC has not already performed the standard command functions, because they (the functions) become the launching pad the IC must use to make such transitions. These IAP changes, many times, must be done very quickly. The command functions are designed to be done from the beginning--in order. It is almost impossible to start doing them in the middle of the operation, particularly when the troops are firmly ensconced well inside the hazard zone, losing the attack battle with the fire, and the smart money (looks at a deteriorating situation) and sez, "It's time to get out of the inside." When this occurs and strong, standard command has not already been established, the IC finds themselves in an absolutely desperate situation and must resort to making spiritual promises (to repent), if the command gods will only somehow extricate the insiders... this one last time. Once the standard command routine gets set up, it's pretty easy to hold (i.e., "shut off") the operational (and organizational) escalation at that level, if the problem gets solved. It's impossible to do the opposite. When the interior fire is getting bigger instead of smaller, and the roof

Command Safety

Review/Revision

Safety Effect:

is coming down instead of staying up, it is a really lousy time for the IC to go back and somehow try to recreate the command functions that should have been put in place twenty minutes ago. This entire book is directed toward the IC who consistently starts to automatically do the strategic level of command every time at every event, no matter how simple and solvable the incident problem looked in the beginning. If the boss of the response system (IC) can't create and maintain such a level of strategic command (which really means doing the command functions), the workers should stay out of the hazard zone--simply, the most beautiful building the capitalists ever built is not worth a firefighter's life. At the end of each chapter (throughout this essay), we have included an overview of what happens when that function is done, and what happens when the function is not done. The detail of those descriptions should create a perspective for the student (us all) that reflects how important connecting the worker's safety and survival is to how the IC does (or does not do) their job. Throughout the entire operation, the command system is the safety system.

Command Safety

Review/Revision

IC Checklist:

☐ Receive, confirm, and evaluate conditions--progress reports.

Safety Effect:

Receiving, processing, and reacting to incoming incident information (in every form) is how the IC stays on top of the ongoing status of people, places, and things. During structural fire fighting incident operations, one of two things happens--the IC's forces are winning or the fire is winning. Based on the welfare of the troops, we must pessimistically concede that all ties are resolved based on the safety of the troops. What this means, simply, is that wherever and whenever the IC determines the fire is going to overpower the size, direction, or location of the attack (BTUs vs GPMs), the IC must quickly reinforce that attack (make it bigger), or move the troops out of the fire's way. Sectors must report back to the IC the progress they are making, as a regular function of managing a sector. This regular information exchange starts with the first and most important tactical priority: rescue. A structure that has been searched and declared "all clear" is now essentially a vacant building (now except for us), and is a different tactical

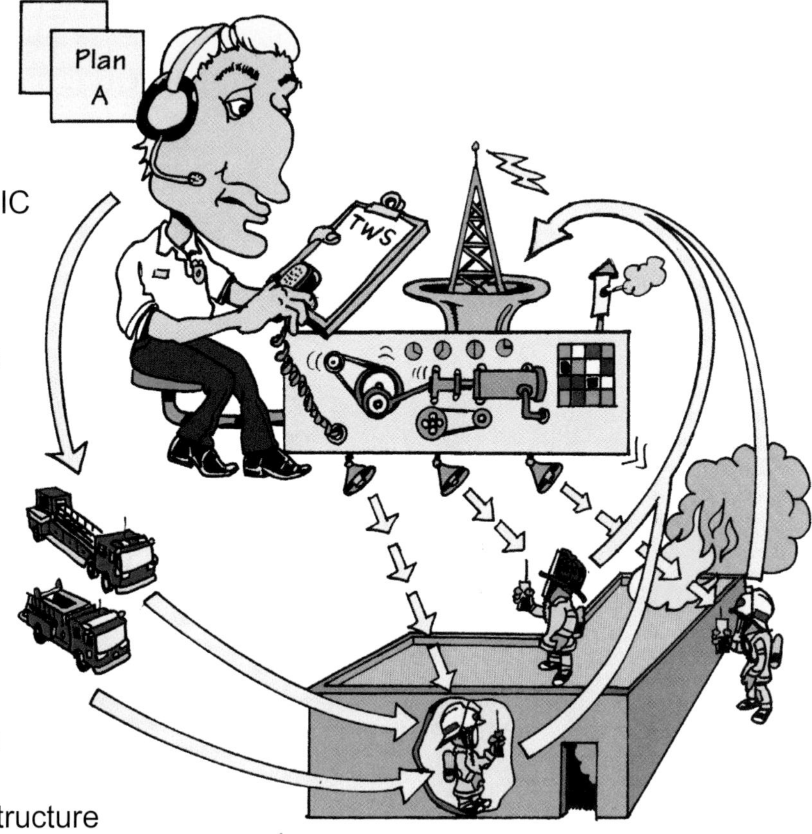

Review/Revision

Safety Effect:

animal than one that hasn't (been searched) in the risk-management equation. This is one of the many (probably the most compelling) reasons for the IC to insist on (and receive) progress/completion reports. If the IC is not receiving these reports, they must ask for what they want to know. Early reports should include progress toward completing the primary search, along with conditions. These reports should start to fill in any information unknowns and "paint a picture" of where the fire (and the building) is now and the direction and avenue where it is all headed (what's left to burn). The IC should base the IAP on the best, current, and forecasted info available, and the IC should let (and help) the troops try to pull the plan off, unless conditions are obviously getting worse. While the IAP should be stable enough so the troops don't feel like a yo-yo, it should always be revisable (and reversible). Progress, completion, and exception reports describe the conditions that keep everyone apprised of how effective the IAP is and identify any pieces of that action plan that need to be fixed. The action plan pieces also allow the IC to keep the plan current and forecast what is going to happen in the future, and then do what they can to support that IAP to make it effective. Everyone operating at the incident must realize that the exchange of critical information about their area/function/safety is an important (and regular) part of their assignment.

Safety Effect:

This "push/pull" information reporting/receiving responsibility automatically goes with every job and is an affirmative function, which means you tell your boss (and they tell you) without being asked when you see, sense or fantasize something that will affect safety and operational effectiveness in any way. Using a combination of face-to-face and radio commo within the regular incident organization creates a quick, simple commo plan that streamlines how information is moved and managed.

Bottom line on info management:

It is the IC's responsibility to seek out information, unknowns, completions, and problems. The IC must realize that sectors and companies may not give reports in a timely manner because they are generally very busy(!) or that time is often distorted inside a burning structure, by the predictable distraction that goes with being directly involved in the action.

Command Safety

Review/Revision

IC Checklist:

☐ Use standard strategy/action plan review items, as the checklist for revision:

- firefighter safety
- does strategy match conditions
- 1/2/3 priority progress
- correct action
- location of attack
- size of attack
- sides
- timing and amount of support
- elapsed operational time
- adequate back up
- adequate resources
- have a plan B (and start developing plan C).

Safety Effect:

After the attack is in place, the IC must engage the standard review questions to evaluate its ongoing effectiveness. Every component of the attack should be monitored and evaluated throughout the event. The number one review item is the safety of the troops. Firefighters routinely (and appropriately... based on the promise we made Mrs. Smith) operate in dangerous positions. It is the IC's ongoing job to improve the dynamic, sometimes quickly changing safety profile of the people (both firefighters and customers), who are in hazardous positions. Many times, this is accomplished by putting the fire out either from the inside (offensively) or from the outside (defensively), based on conditions. If the fire doesn't react according to plan, the IC must decide if the current strategy still matches the conditions.

Command Safety

Review/Revision

Safety Effect:

Progress reports will weigh heavily in this decision. If the current strategy is still valid, then the IC must adjust, support, and reinforce (if needed) the attack in such a way that it overwhelms the fire and makes it go bye-bye. One of the things that should be "idling" in every fire attack is the ongoing, standby safety-recovery system that is in place to provide for the quick and effective rescue of firefighters, when necessary. The regular command and operational system must have a built in (RIC team[s]) rescue component. The smart front end of this recovery system must be automatically put in place, whenever there is a hazard zone present. If the initial attack hit didn't solve the problem (put the fire out) and conditions are worsening, it is many times a better tactic to pull everyone outside, and discuss plan B over a paper cup of Gatorade and a cold Big Mac. The dangerous "other side" of the evaluation/reaction process is staying in untenable positions too long, and having to send rescue crews in to extricate (our) people. The IC has got to maintain an "on-deck" tactical reserve (RIC teams) to do firefighter extrication, when something unexpected happens. These sudden, breathtaking, unexpected, and often very harmful events sometimes do occur, even if all the system pieces are effectively in place (this is what makes being a firefighter different from running a well-maintained drill press or working in a nice quiet office), and the IC needs to always be ready to quickly react to such changes. On the other hand, if the fire is getting bigger and all the tactical evaluation news has been bad, the IC needs to

Command Safety

Review/Revision

Safety Effect:

proactively pull the plug on offensive operations, and switch to a defensive strategy. This is known as smart management, and the following basic review questions become the foundation of that proactive approach:

☐ Firefighter safety
Has the IC provided for the safety and welfare of the firefighters?

Strategy
Conditions

☐ Fire stage/strategy
Does the overall offensive/defensive strategy match the current conditions?

☐ Adequate resources
Has the IC called for enough resources to overcome tactical problems?

☐ Have a plan B (and C, and D, and...)
Has the IC formulated a major attack plan revision?

☐ 1-2-3 Priority progress
Is the operation moving along the basic order of rescue, fire control, and property conservation? Does the risk level match the current priority?

☐ Correct action
Are fire fighting crews/sectors applying the correct (effective) operational action (techniques, evolutions, procedures) on the tactical and task level?

☐ Location of attack
Are the correct key attack positions being covered?

Command Safety

Review/Revision

Safety Effect:

- ☐ Size of attack
  Is the attack large enough to control the fire?

- ☐ Covering the sides
  Are we aware of conditions on the top, bottom, inside, four sides, and the layers and concealed spaces that are attached to those sides?

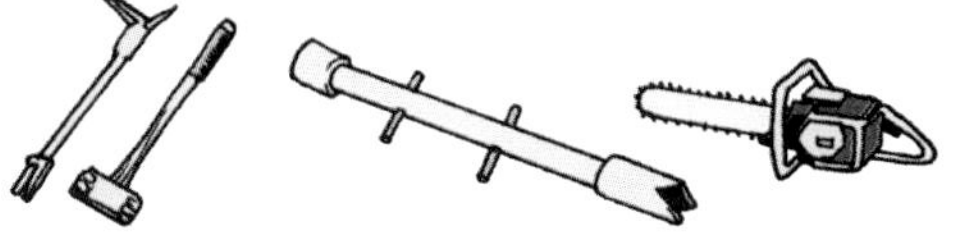

- ☐ Timing and amount of support
  Is the fire attack receiving adequate and well-timed support (ventilation, forcible entry, access provision)?

- ☐ Effect of operational action
  Is the attack working--are conditions getting better/unchanged/worse?

- ☐ Elapsed operational time
  Is time running out on offensive attack positions?

- ☐ Adequate back up
  Are resources in place to reinforce the current operation and are RIC teams in place?

- ☐ Operational control
  Does the IC have effective command and control of the operation?

Command Safety

Review/Revision

IC Checklist:

☐ Quickly make transitions based on the safety profile of changing/forecasted conditions:

- visual/gut stuff
- feedback from sectors/companies
- quick evaluation
- move the troops
- regroup--go to plan B.

Safety Effect:

As we have discussed (and discussed and discussed...), our response system thankfully has a very strong natural inclination to act aggressively. Based on their inherent attack energy, it's pretty easy to get and keep our troops in attack positions where they can directly engage the incident problem. While this attack inclination is the strength of our service, if it's not effectively managed and controlled, it can also become a major weakness (i.e., taking a knife to a fun fight). Situations that require the IC to quickly move the attacking troops (for whatever reason) will define, on the most basic level, if the IC is in or out of control--a major objective of the eight command functions is to create a management system that is always able to control the position and function of the hazard-zone workers. The IC must evaluate

Safety Effect:

feedback from all over the incident site to add up the safety score--when that score goes negative, the IC must move the troops to a safe position, get a PAR, and go to plan B. What going negative on the safety score means is that the incident hazard is bigger than the safety system response. Maintaining an ongoing awareness of that score is a major safety responsibility of the IC. Up to a certain point, the IC can increase the safety response by calling more responders and by bumping up the standard parts of the safety system. Calling more resources, including working companies, safety officers, increasing the RIC capability, adding more command officer-level sector bosses, increasing the backup lines, and support to the interior attack are all ways the IC can make the safety system bigger. Beyond a certain point, an escalating incident, with a lot of property left to burn, can simply get bigger, quicker than the local resource (deployment) capability of the IC to escalate the operational and safety response. The IC must maintain control of the transitions (i.e., IAP revisions) that are required to "cooperate with the inevitable." Being able to effectively make these transitions requires the IC to establish strong command from the very beginning of operations (the actual execution of the eight standard command functions = strong command). A major piece of the safety system involves the command team using the regular command functions as the foundation for operations. This means that responders go to their assigned positions and perform their assigned functions under the command and control of the incident

Command Safety

Review/Revision

Safety Effect:

organization (IC and sectors). Operations are established based on SOPs, direct orders, and the conscious decisions of officers who exchange current (and forecasted) information on their location, work progress, and welfare. Everyone operates with the understanding that the IAP will continually be reviewed (and revised if necessary) based on the "safety score" and that every part of the organization must participate in this review. Every operating and command position has its own unique perspective on current conditions and safety, and to be accurate, the overall safety score must be the result of information from all over the incident being received, processed, evaluated, and then managed by the IC for the overall incident operation. This is a major reason we establish a strategic-level sitting boss (IC), as quickly as possible. When the safety score goes negative, the IC must use the same command functions to get the workers out of the hazard zone that were used to get them into that zone (strong initial command, control of the commo process, specific orders to specific companies, standard resource tracking/quick organizational development, and utilization, etc.). How the IC sets up the incident in the beginning pays off (big time) in the middle, when based on changes, they must now quickly move the troops away from dangerous conditions. If

Command Safety

Review/Revision

Safety Effect:

everyone is in their assigned place, effectively connected to the IAP, and mutually processing up-to-date, accurate information, moving the troops can be done in a smooth and quick way. If a lot of troops free-lanced their way into a lot of different and unknown (except to them) places all over the hazard zone, they are basically disconnected from any command protection, so it's pretty tough for them all to free-lance their way out in a safe, timely, coordinated, and orderly way. It's lots better to start under control, stay under control, and never lose control.

NOTE:

Sometimes all the standard command and operational stuff is set up and in place, but something just doesn't feel right to the IC. In these cases, the IC's gut is sending the brain a message that some part of what is going on doesn't fit the normal profile or pattern for that situation. Experienced ICs acknowledge and react to such messages, particularly where the troop's safety is concerned. When

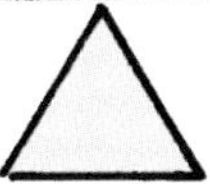

Command Safety

Review/Revision

Safety Effect:

this occurs, it ain't voodoo--it's something that's occurring in a subtle way that is just below the conscious level that is blinking "red flag" in that special corner of the IC's brain that is set up to process such stuff... smart ICs pay attention to their whole brain.

Command Safety

Review/Revision

IC Checklist:

- ☐ Quickly provide "Salvage Command," if necessary.

Safety Effect:

Safety related systems, procedures, and practices are designed to prevent us from injury and death ***while*** we are delivering service in the hazard zone. The safety responsibilities and routines for the three organizational levels (strategic, tactical, and task) are pretty straightforward and designed for each level to follow the rules and operate within their role. The IC is responsible for everyone's overall safety. The IC uses the eight functions of command to achieve the completion of the tactical priorities and manage incident scene safety for all the incident players.

Offensive fire fights are typically our most hazardous tactical undertaking. A fast attacking company officer IC usually kicks off these high-adrenaline festivals operating as a triple-duty player (strategic, tactical, and task), who attempts to deliver enough action and command to solve the incident problem and eliminate the hazard. This is how the system is designed. The system also has a built in

Command Safety

Review/Revision

Safety Effect:

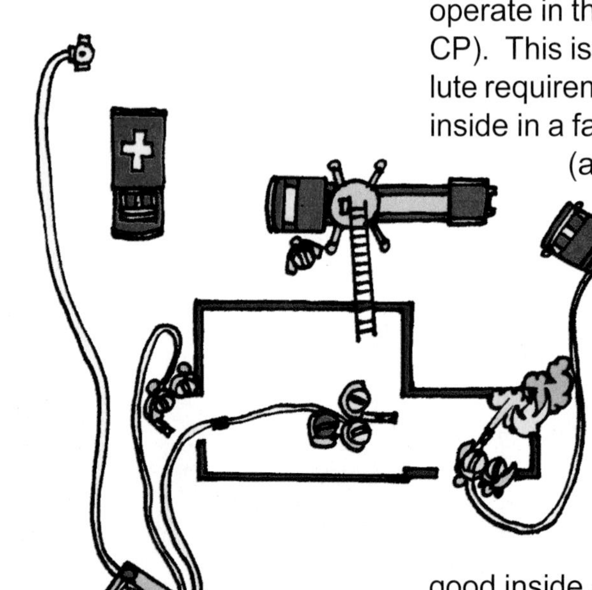

evolution/escalation for cases where the initial wave of problem solvers isn't able to eliminate the hazard. In these cases, command typically gets transferred from a fast-attacking company officer to an IC who will operate in the command mode (inside a CP). This is not a suggestion--it is an absolute requirement. Many times, IC #1 is on the inside in a fast-attack position, so it is difficult (and dumb) for the arriving officer (who will become IC #2) to go into the hazard zone to do a face-to-face command transfer. The objective of command transfer is to upgrade command, so there is little point to shifting from one mobile IC, who is in a disadvantaged command position (but is in a good inside attack position), to another mobile IC who would be in essentially in that same disadvantaged (from a command standpoint) position. The standard (eight) command functions provide the basic outline for IC action. How these functions are being accomplished forms the checklist for command effectiveness, at any time in the operation. How the current IC is doing (with the functions) becomes evident when the command transfer process occurs. Simply, the command transfer recipient (IC #2) will become quickly (and sometimes painfully) aware of how the donor (IC #1) has set up and performed the command functions up to that point, based on what the recipient inherits--or in some cases if there is, in fact, anything in place to pass on to the next IC.

Command Safety

Review/Revision

Safety Effect:

While a fast-attacking IC can quickly deliver a quickly organized response, that can often times solve the incident problem, they cannot manage the overall strategic safety requirements for a rapidly expanding offensive fire fight. When one or two well-placed attack lines do not control a fire inside a structure, it is a sign ("red flag") that the fire poses an elevated risk to our inside operators. We will discuss the safety nuts and bolts of the transfer and continuation of command, in the next enlightening chapter. We only mention it in this chapter because it is not unheard of for a later-arriving IC (IC #2) to show up at an incident where the first IC has failed to do the functions of command. It really doesn't matter very much why command got out of whack when IC #2 arrives on the scene. Before the second-arriving IC can have an in-depth, "philosophical" analysis with IC #1 regarding incident management, IC #2 has got to somehow quickly make sense of what's going on, figure out where everyone is, determine if it is safe to continue with the current operation, and get all of the kittens back into the box. This is best described as having to salvage command.

When one firefighter commits an unsafe act, they put themselves and those around them

Command Safety

REVIEW/REVISION

Safety Effect:

at a greater risk. When the IC allows the IMS to become out of balance and ceases to manage strategic-level safety, all of the hazard-zone workers are exposed to a much greater risk. When this occurs, the next-in IC must aggressively take control of the incident operation (i.e., "salvage command"). The transferring IC will have to conduct a fast-and-dirty "shorthand," using the regular eight functions of command to get the operation back under control. Being able to quickly implement an effective level of incident control is the absolute top of the incident command food chain in maintaining hazard-zone worker safety and survival.

The current IC must always do the regular command functions, as long as there is a hazard zone present. There is no place in the standard command routine where the IC completes the regular command functions, and then exits into another management mode, and now engages in some kind of cosmic command(?). The IC must always be doing the regular command functions throughout the beginning, middle, and end of the event. No one will know more about how well the current IC has (or has not) set up command than the person (i.e., the "next" IC) who inherits IC #1's operation. How the last IC has done those functions will determine the action the next IC must take, particularly in the beginning period of taking command. When a response boss arrives at an incident

Safety Effect:

where hazard-zone action is underway and regular command is not adequately in place, that person must take command by doing a set of things to establish (what is in effect) midpoint command. This "salvage" action can range from doing a minor tune up in a well-executed, fast-attack situation to a major overhaul, to somehow get an extensive free-enterprise beginning under command and control. The IC's ability (or lack of) to salvage command could possibly be the decisive factor on whether firefighters return home unharmed. The regular command functions serve as the foundation for this revision and the following would include some of the (command function) action involved for the person who will become the next IC, and must perform "salvage command."

Function #1
Assume, Confirm, and Position Command

Function #8
Continue, Transfer, and Terminate Command

When we talk about instances where IC #2 has to salvage command, these two functions (one and eight) naturally end up coming together. Situations where command must be "salvaged" are the product of the initial IC somehow letting it get out of whack. After IC #2 arrives at the scene, they must take charge (transfer command) and effectively begin managing

Safety Effect:

the other seven functions. This process begins with contacting the initial IC (generally over the tactical radio channel) and getting a progress report. This process is dependent on having an IC in place when IC #2 shows up at the scene. In cases where multiple units are operating and no one is in command, IC #2 not only has to start from scratch, but also has to bring under control what amounts to a collection of free-lancers. No IC equals no command and control effectively in place. The result of this lack of command and control produces essentially unprotected hazard-zone workers (and customers).

Simply upgrading the IC into a command position (inside a vehicle) raises incident scene safety several notches. The IC is now in a position where they can actually manage everyone's safety on the strategic level. This also places the IC in a standard command position where they can implement and manage an incident organization (sectors), that is responsible for tactical-level safety. The only shot we have at getting an out-of-control operation back under control is to put the IC in the most effective command position possible.

Function #2
Situation Evaluation

The IC uses situation evaluation to determine the correct strategy and corresponding IAP. Many out-of-control incident situations are

Safety Effect:

the result of the initial IC skipping this function. In our enthusiasm to engage the incident problem, it is very easy to fall into the ready-fire-aim syndrome. One of the side effects of blind aggression is, when we finally roll our eyes back into their proper focus, we often find ourselves in locations and conditions that are no longer tenable. This puts us in a position, where if we don't eliminate the incident problem, we buy the farm. A proper incident size up provides the IC the information to decide if the incident conditions are controllable and survivable, along with the best task-level action required to solve the problem (i.e., IAP).

When the initial IC fails to do an adequate size up, IC #2 has to quickly determine what the most critical incident factors are--particularly the ones that affect safety. For structure fires, this will always include the size of the fire, the intensity of the products of combustion, how long the fire has been burning, and what is on fire (contents versus structure, highly volatile fuels, etc.). The IC determines the scope of the incident problem by comparing what they see from the command post with progress reports from companies/ sectors, that are operating in hazard-zone positions and in areas that the IC cannot see from the CP (i.e., roof, rear of the structure, etc.).

The IC has to quickly find out what's going on in the key places that affect worker safety. In most structure fires, these places are the concealed spaces, particularly

Safety Effect:

attics, basements, and any voids that the fire can use to quickly take possession of large parts of the building. Stairs, halls, lobbies, and "public places" also are critical spots because they create the normal access areas where the fire, the firefighters, and the customers can move inside the structure. Regaining control of an out-of-balance structure fire, that can be controlled using the offensive strategy and several well-placed attack lines, is a safe and righteous venture. The same basic situation, with a deep-seated fire in the attic of the building, is a completely different animal (time to change strategies). The IC uses Function #2 to figure out the incident conditions so they can match operational action (and consequences) to those conditions.

Function #3
Communications

The basic tool the IC has at their disposal to "salvage" command is the radio, a TWS, and command helpers/team. In the old days, when the IC showed up to an out-of-control event (many opportunities then, as most of our events where fast and furious), they would promptly launch themselves into the hazard zone, adding another layer of confusion to an already confused event. It is impossible to get control of an unmanaged situation, if the IC is in a lousy communications position (out of the vehicle). The IC must be the central incident scene communications player. The

Safety Effect:

strategic level brings everything together, including the overall scene safety plan, by connecting all the separate tactical and task units together over the tactical radio channel.

The IC uses the radio to determine the current status of all of the hazard-zone workers. The IC must quickly determine who is in the hazard zone, what they are doing, are they okay, and do they have a PAR. This initial batch of communications furnishes the IC with most of the safety and operational info they will need to begin the process of getting the incident under control. These initial reports from operating units should also include the conditions in which they are operating (situation evaluation), and any request for any additional resources that are required (deployment). The IC should also begin to assign sector-officer responsibilities to the officers of these initially-placed companies (organization).

The IC uses the radio and what they can see from the CP to do the eight standard functions of command. This becomes a major part of the system that the IC uses to bring the unorganized, free-lanced beginning back under control.

Function #4
Deployment

Level-one and level-two staging and assignment by the IC are two of the major IMS mechanisms we use to maintain control of incident scene resources and personnel. They also serve as major pieces of our

Safety Effect:

safety system. When IC #2 takes over an out-of-control event, it can be the result of IC #1 not taking command. Not having an effective IC in place can cause later-arriving units to free-lance their way into the hazard zone. Free-lancers rarely announce the action they are going to take over the tactical channel. This puts the later-arriving IC #2 at a severe disadvantage. IC #2 may be under the impression that only two or three units are operating in the hazard zone when, in fact, there are four or five. The IC must then quickly determine who is operating in the hazard zone, if they are okay, how long they have been in their current position, and how much longer they can stay there (i.e., how much air they have left), who is staged and available for assignment, who is still responding to the scene, and if there are enough resources assigned to the incident.

The IC must decide if the operating resources are in correct/safe positions. If they aren't, this will require more resources, so the proper position can be covered. In cases where everyone is acting like there is a full moon and nothing is going right, it may be easier (and safer) to pull everyone out, get PARs, strike another alarm, and start over.

Function #5
Strategy and Incident Action Planning

Function #7
Evaluate, Review, and Revise

When IC #2 finds themselves in the difficult position where they have to salvage command, Functions #5 and #7 become closely

Safety Effect:

connected, and must be addressed together. Function #5 must be completed and shared, as a standard practice with all of the other responders, as part of the IC's (IC #1) IRR. When this doesn't happen, IC #2 has to guess what incident conditions looked like when IC #1 arrived on the scene, what effect the control efforts have had in the ensuing time, and if units are operating in safe and supportable positions.

This function is a major factor in the overall incident scene safety picture. In situations, where there are three or four companies operating in the hazard zone, the IC okays/validates/permits/endorses/supports that operation, and allows firefighters to remain in those positions through strategy management (in this case, remaining in the offensive strategy). When conditions take a turn for the worse and move from offensive conditions to defensive ones, the IC must then change the overall incident strategy, pull everyone out of the hazard zone, and change the IAP to reflect that strategic change (from offensive to defensive). This simple act prevents the incident hazard from injuring or killing the hazard-zone workers.

In situations where IC #2 shows up to the scene and has to take command of a previously unmanaged incident (shame on IC #1), the first thing they must determine is if the hazard-zone workers are operating in the correct strategy. Most unmanaged events start off with the initial units jumping (or

Safety Effect:

blasting themselves) into offensive positions. In many cases, these are the correct positions and some other part of the incident management system is out of balance (no incident organization, no overall IAP, etc.) and IC #2 has to do some quick command work to get the incident back on track, keep the operation moving forward, and keep everyone safe. In other cases, if the hazard-zone workers are operating in offensive positions under defensive conditions, it is time to pull the plug, move everyone to safe, outside locations, get PARs on all the hazard-zone workers, and change the IAP to match up with the new defensive strategy.

After IC #2 confirms that the incident operation is taking place within the proper strategy, it is time to evaluate the current plan, making sure that the proper positions and functions are covered. Most out-of-control incidents lack a well-thought-out and managed IAP. The IC (and everyone else) makes things safer by eliminating the incident problem (i.e., putting out the fire). Using a shotgun approach, and simply throwing resources at the incident problem is not a real plan. These "mass-blast" festivals rely on some company stumbling into the right place and applying the correct action. This approach may work when the incident problem, the building where the problem is taking place, and required level of resources are small enough that somebody can quickly engage and overwhelm the problem. When the incident problem, the building size, or the number resources are beyond a certain

Safety Effect:

size, this essentially unmanaged free-lancing approach becomes very dangerous and reckless.

The core of the IC's plan must always include the safety and welfare of the hazard-zone workers. If the scene becomes a safer place for the workers, it will also be a safer place for the customers.

Function #6
Organization

Developing and using a sector-based incident organization gives the IC the capability to quickly decentralize command and control. Placing supervision and management directly into the hazard zone creates an enormous safety improvement. Not only does the IC push tactical-level supervision and management where it belongs, they also establish communications partners in the key tactical/operational/hazard areas. This helps to streamline the overall communications process (maximizing the available air time), places reporting agents in all the key spots, and places critical-tactical managers in the hazard areas.

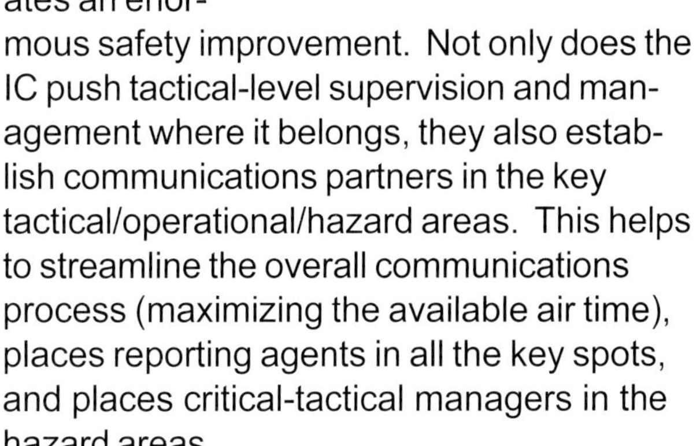

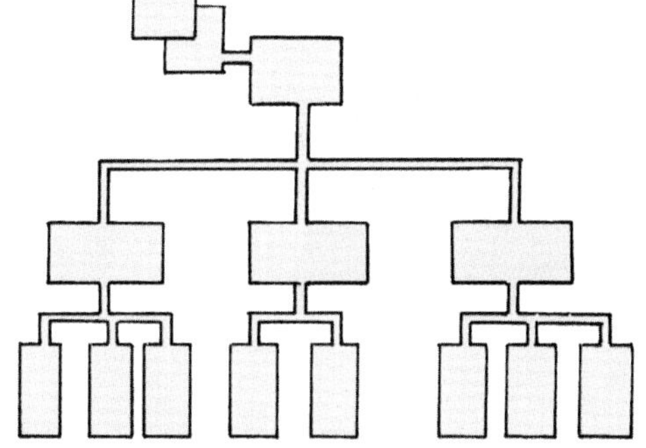

Creating an effective incident organization is the quickest way to bring a runaway incident

Safety Effect:

operation back under control. During the initial stages of incident operations, the strategy and IAP describe how the IC is going to solve the incident problem. The strategy and IAP get executed (and begin to solve the incident problem) when the IC shares it with the rest of the incident organization. This process is also where we begin to address overall (strategic) scene safety. Ongoing incident scene safety is largely managed by sector officers who constantly evaluate and report changing incident/sector conditions, and how they react to the control efforts.

The IC maintains their ability to manage on the strategic level by delegating management and supervision responsibility to sector officers. This also has the safety effect of putting management partners directly where the work is being performed. A problem can arise when the IC assigns too much resource to a sector officer who has a lot of other stuff going on in their life. Traditionally, initial-sector officer responsibilities are assigned to the first unit/officer assigned to that area. This gets the incident organization up and running. As quickly as possible, the IC must upgrade these positions to later-arriving command officers.

Command Safety

Review/Revision

Safety Effect:

While a company officer can get the sector operation started, they are not in a position (nor is it their role) to manage a quickly growing hazard-zone sector. In many cases, the company officer is responsible for managing their own crew, assisting with task-level activities, and are not in a very good management position. As the resource requirements in a sector increase, the sector officer position must be upgraded to match the management/safety requirements.

Command Safety

Review/Revision

This happens when Revision is done:

- ☐ IC #1 starts the standard command functions in their proper order at the beginning of operations to control ops.
- ☐ IC #1 develops strategy/IAP based on best possible information.
- ☐ IC develops organization to match incident profile and resources.
- ☐ IC uses ongoing CP visual and decentralized incident organizational reporting to continue and strengthen ongoing evaluation.
- ☐ Command team is continually evaluating operational effectiveness and discussing if plan A should evolve into plan B.
- ☐ IC revises strategy and IAP to match work progress, changing conditions, and to ensure tactical benchmarks are met.
- ☐ Troops are always protected by up-to-date, risk-management based IAP.
- ☐ Customer service is performed, based on completed objectives.
- ☐ Standard risk management is reflected in decision making.

Command Safety

Review/Revision

This happens when Revision is not done:

- ☐ Standard effective command functions are not done in the front end as a base for revision.
- ☐ Task-level resources are increased (i.e., called for) and command level capability is not increased (command development starts and stays behind the incident-power curve).
- ☐ The IC does not effectively forecast changing conditions.
- ☐ IC doesn't trust or listen to the troops--bases plan on incorrect/outdated/preconceived information.
- ☐ No effective organizational reporting going on.
- ☐ When conditions change, IAP becomes obsolete:
  - changes are now driving the IC
  - IC tries to play catch up
  - resource allocation is out of whack
  - IC can't develop IAP plan B quick enough
  - problem keeps winning--IC keeps losing.
- ☐ Firefighters end up in offensive positions at defensive fires.
- ☐ IC loses control of hazard-zone operations.
- ☐ Customers/troops get beaten up (or worse)--savable property is trashed.

Command Safety

Review/Revision

COACHING VERSION

"Set up command in the beginning, so you can move and maintain control later on, and always match your current plan with current conditions. Pay attention and keep forecasting where the problems are going. Remember that all fire stages are not created equally, so all IAPs must be developed to match the actual incident profile. The fire can stay at one stage, and then conditions can change very quickly. You must always be reading the "gauges" to forecast (and react to) what will be going on in the next five/ten/fifteen minutes. Use strong sector and ongoing communications feedback to stay mean and mobile. Pay attention and quickly react to anything or anybody that is causing you to be uncomfortable. Manage conditions by controlling the troops and their action--don't live with a bad situation... be careful of doing the wrong things harder... don't stay in or reinforce bad (unsafe) positions... evaluate, react, revise, do something differently to get ahead of the incident problem. Always maintain a way out of the hazard zone that is protected with attack lines and crews, so firefighters are in a retrievable position. Don't fall in love with the IAP... stay agile. Always consider retreat and survival (as an option), and be ready to move quickly. Unless there is a possible critical rescue, be extremely careful of having firefighters positioned so close to desperate fire-control situations, that the only way they can possibly survive is to put the fire out. Don't screw around with deteriorating conditions. Don't do work you're uncomfortable with. Establish a plan, and then keep it current--remember, it's almost impossible to revise a "non-plan." If you arrive as IC #2, evaluate how command has been set up, and develop whatever is required to fill in any "holes." If what is in place is working, support it; if it is wrong/unsafe, fix it; if it is not in place, make it happen. You must do it all in a strong, quick, supportive way.

Command Safety

Continue/Transfer/Terminate Command

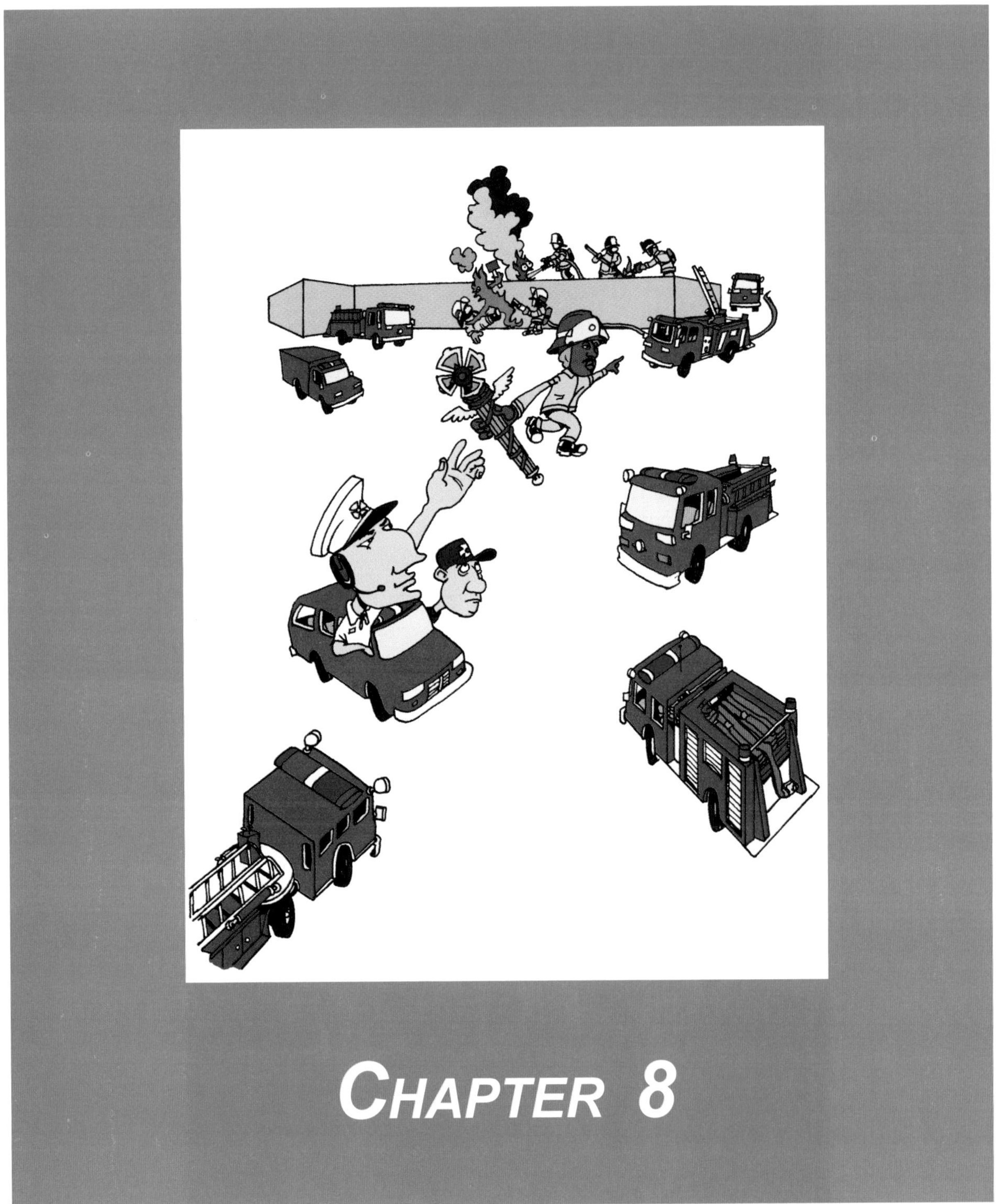

Chapter 8

Command Safety

Continue/Transfer/Terminate Command

Command Safety

Continue/Transfer/Terminate Command

Function 8

Continue/Transfer/Terminate Command

Major Goal

To develop a standard approach to command transfer and to operating the midpoint and final stages of command.

IC Checklist:

- ☐ Estimate the length of command required.
- ☐ Consider the time for completing each tactical priority.
- ☐ Consider life safety, fire area profile, and fire conditions to estimate incident duration.
- ☐ Develop and support an organization that outlasts the major problem(s).
- ☐ Assume, maintain, and upgrade effective command positioning.
- ☐ Develop and maintain effective fireground communications.
- ☐ Keep the attack plan going.
- ☐ Use standard command transfer (both ways).
- ☐ Exchange all pertinent information between IC and company/ sector officers.
- ☐ Place resources back in service with a demobilization plan.
- ☐ Insure adequate critique process is underway before decommitment.
- ☐ Provide required critical incident support.

Command Safety

Continue/Transfer/Terminate Command

IC Checklist:

☐ Estimate the length of command required.

Safety Effect:

The command system must establish and maintain itself for a length of time that out lasts the incident problem. If this does not occur, the incident simply "burns" past (i.e., out performs) our command system and we lose our capability to take control of incident conditions, so we can protect the customers (and their stuff) and our troops. Quickly forecasting the future and calling for adequate resources is how we reinforce and strengthen our operational capability. Assuming a CP position, creating a command team with an IC, SO, and SA, and expanding the assignment of geographic and functional sectors are ways we reinforce and strengthen command. Command is transferred from the current IC to the next IC based on the arrival of a higher-ranking officer, someone with a special capability, a duty officer, a command team member, someone who arrived in a "better" vehicle, etc. If the initial IC is a company officer and the problem is going to be over in ten minutes, and the next responder who is next in the IC "food chain" is nine minutes out, it is probably unnecessary to transfer command (unless the last minute of the active incident is going to be really exciting). In the ten-minute (small-sized incident) case, it doesn't make much sense to call for a big-time command support response based on the predicted short period of operations. The length (of time) of

Safety Effect:

the incident will determine the amount of operational resources required to control the problem. Ten-minute problems can be very intense and dangerous and require effective, alert, quick, up-front command. We must understand that wherever we must operate very quickly, we lose the discretionary time advantage and this creates an increased level of danger to our troops. These quick incidents are many times handled with one or two companies. Hour-long problems are more "time forgiving" and will generally require more resources, along with the necessary command components required to manage them. The entire command/safety system must understand and manage the finite energy capabilities of the entire response system. Companies (like virtually everybody and everything else) have limits to the amount of work they can accomplish at an incident. This is also true for the IC. A person (even a well-supported one) can be ground down to an ineffective level after being the IC for too long. While it may appear (to the uninitiated) that the IC is just sitting in the CP quietly (we hope) talking on the radio, and making notes on a TWS, we must realize that command work is intense and exhausting. The system must create a realistic understanding of the dynamics of command fatigue and develop a system that automatically provides for relief for the command team. We must provide the same rotation and rehab for the command team, as we do for the hose haulers, and the ceiling pullers. If the incident is going to go beyond a short-term

Safety Effect:

event, some type of IC rotation schedule must be automatically implemented. The IC (and the command team) should be rotated, given a break, and "rehabbed" at certain intervals, especially during intense operations. The goal is to keep from burning out (pun) the entire command staff during the first two hours of a four-hour incident. Providing an IC who is lucid and conscious becomes a major safety consideration in protecting the hazard-zone workers. The Federal Aviation Administration (FAA) requires an effective work/rest mixture for the cockpit crew that is flying Aunt Mildred through the friendly skies, back to homeburg. We should require the same for the character (IC/command team) who is responsible for E-1's hazard-zone in/out trip. A good management model uses a one-hour threshold for the planning process to prepare the incident response for a longer duration event. Be careful of having an incident "creep" in small-time increments to the point that it burns out the unrotated/unrehabbed workers and the command team. At one hour, we should prepare formally for whatever is next. It's lots better to recognize the length of operations

Safety Effect:

sooner and to develop a realistic plan to match that projection, than to have to rush to recruit a relief command team after the current players become so worn out they start babbling on the tactical channel and eventually fall out of their seats(!).

Anyone who is operating (anywhere) on the incident site beyond their effective energy level is a hazard to themselves and everyone around them. A major IMS objective is to continually evaluate operational and command effectiveness, and a big part of that capability involves maintaining an awareness of the ongoing condition of the humans operating on every level. Our responders will naturally push their limits based on their dedication and commitment. Bosses (on every level) must continually evaluate where everyone (including themselves) operating on the incident is on the energy scale. This also applies to the SA who must create an effective plan for command-staff relief. Many times, the IC will develop the feeling that the incident is "their fire" and that they must be the ongoing, eternal "nobody is going to relieve me" incident commander--such personal ownership makes it difficult for that person to objectively disconnect and let a fresh player take over. The SA must control this admirable (yet dysfunctional) approach, and must make it clear that the incident "belongs" to the entire response team and that command rotation is not optional.

Command Safety

Continue/Transfer/Terminate Command

IC Checklist:

- ☐ Consider the time for completing each tactical priority.

Safety Effect:

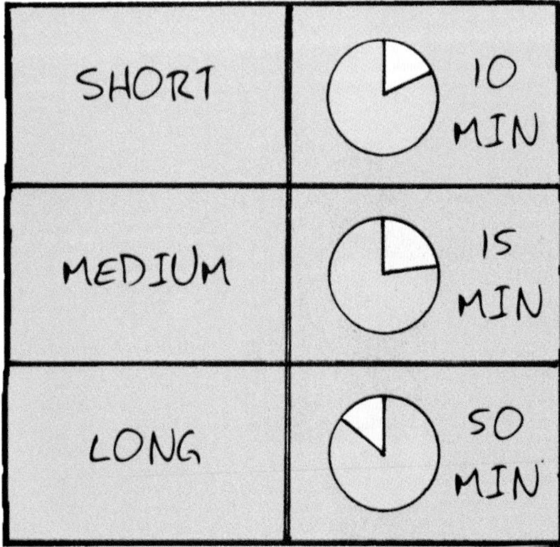

Sometimes breaking the incident down into smaller operational/planning pieces will give the IC a better idea of how long the entire operation will last, and what those different phases will look like. Using the standard rescue/fire control/property conservation tactical priorities as the framework for incident action will better define the high-risk portion of the fire attack and how long it will last. Connecting the regular big/little/no-risk management plan levels to the three standard operational tactical priorities also helps everyone understand the standard operational approach. If the IC forecasts that their forces will have an "all clear" and then an "under control" in ten minutes, this becomes a management barometer on the projected progress for those objectives. If ten minutes later the fire is still booming, it is time to start making some serious overall strategy/operating position safety decisions. Through training and ongoing experience, officers and crews should be taught that if normal task-time lines are interrupted or exceeded, it is mandatory that the IC be informed. When estimated standard times are exceeded, they should be reported using condition/progress/exception reports, and if these reports relate to hazard-zone safety, the IC must regard these "time distortions" from normal times as safety "red flags," and respond to them as such. The IC needs to

Command Safety

Continue/Transfer/Terminate Command

Safety Effect:

forecast ahead of current conditions, but must always focus on what is happening at the current time. Don't get bogged down worrying a whole lot about salvage operations, if crews are having trouble searching the building. A big-time place in the standard evolution of an incident occurs in between rescue and fire control. This is where the system says we shift from a big (rescue) risk to a little (fire control) risk--if the IC gets an "all clear" and there is a big risk still present, the IC must go defensive (this is why we call it risk management). Once the fire is controlled, activities will slow down and safety will be much easier to manage, and companies can provide a better size up of what it will take to wrap up (complete) the operation. The three-part (tactical priority based) forecast will help more clearly define the resource, command, and safety needs for the incident.

Command Safety

CONTINUE/TRANSFER/TERMINATE COMMAND

IC Checklist:

☐ Consider life safety, fire area profile, and fire conditions to estimate incident duration.

Safety Effect:

The IC must evaluate the critical incident factors to determine how long the operation will last. This evaluation process begins with the initial size up and continues throughout the incident. One of the decisions that must emerge from this ongoing evaluation is for the IC to create a command organization (and an operational response) that matches the incident nature, size, and complexity, and is initially set up quickly enough to effectively begin to intervene in the incident problem. The organization must then grow and reinforce itself fast enough to out last the hazards the fire is producing. This is a major way the IC protects the workers. Evaluating the life-safety (rescue) profile of the event defines the high-hazard inside work the response team must do. The number, location, and condition of the customers possibly trapped in the hazard zone becomes a critical part of the IAP. Evaluating the fire area profile describes the barriers and challenges that are present and

Safety Effect:

becomes a major part of the support plan (forcible entry, access, ventilation, lighting, etc.). The IC must put in place an IAP to create the capability for the troops to get inside, and to support them in those interior positions, so they can get the job done. Evaluating and responding to fire conditions is how the IC reacts to what is actually causing the incident to be an incident. The IC must create an operational response to find the fire, cut it off, and put it out. The IC uses the consideration of these critical factors as the very practical basis to decide how much command will be required and how long that command must go on. The faster the IC creates this command system capability, the faster our organizational response can get ahead and stay ahead of the incident problem(s), and the safer the operation will be... sadly, the opposite is also true. Playing command catch up always creates an increased level of danger for hazard-zone workers.

Command Safety

Continue/Transfer/Terminate Command

IC Checklist:

☐ Develop and support an organization that out lasts the major problem(s).

Safety Effect:

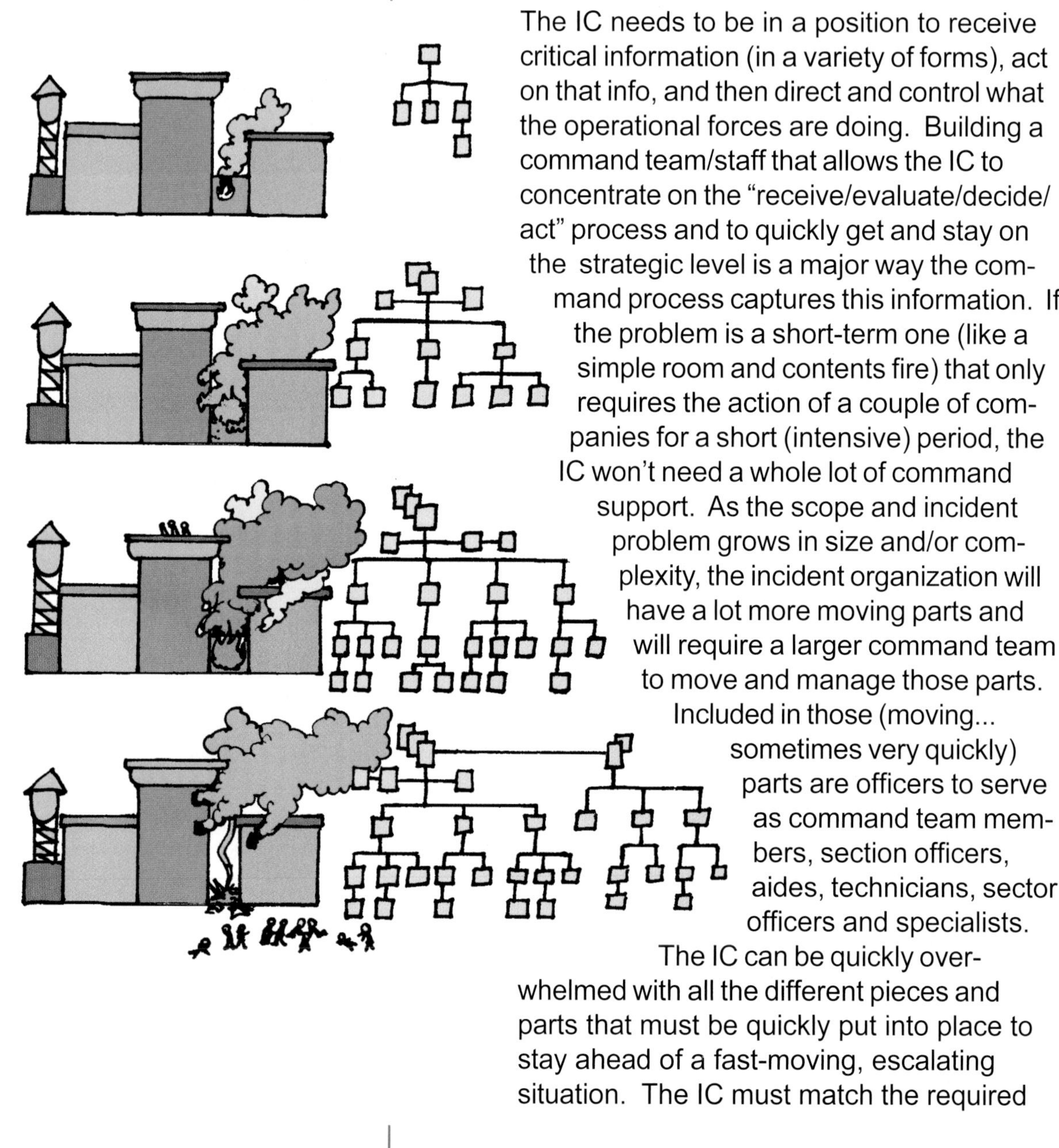

The IC needs to be in a position to receive critical information (in a variety of forms), act on that info, and then direct and control what the operational forces are doing. Building a command team/staff that allows the IC to concentrate on the "receive/evaluate/decide/act" process and to quickly get and stay on the strategic level is a major way the command process captures this information. If the problem is a short-term one (like a simple room and contents fire) that only requires the action of a couple of companies for a short (intensive) period, the IC won't need a whole lot of command support. As the scope and incident problem grows in size and/or complexity, the incident organization will have a lot more moving parts and will require a larger command team to move and manage those parts. Included in those (moving... sometimes very quickly) parts are officers to serve as command team members, section officers, aides, technicians, sector officers and specialists. The IC can be quickly overwhelmed with all the different pieces and parts that must be quickly put into place to stay ahead of a fast-moving, escalating situation. The IC must match the required

Command Safety

Continue/Transfer/Terminate Command

Safety Effect:

command support to the forecasted size of the command organization for the length of time the operation will last. This includes calling a sufficient number of players to provide relief and support for sector officers and command staff. It's lots easier for the IC to send home a group of officers, who were quickly called to build a command team to manage what could have been (but didn't develop into) a big deal, than to play command catch up when a tactical situation sucker punched and surprised a lonely, half-asleep IC. These are the exciting events where the firefighters must jump out of windows (as an example) to save themselves because the overmatched IC lost control of the inside of the building. The organization has to be big enough to manage the deployment of required resources, and to maintain command and control long enough to outlast the incident problem. The command system must develop procedures and techniques to provide rehab, rotation, and relief for the IC and the command team during extended operations. The IC must estimate how long the event will last, and initiate the response of those support resources. It's lots better if this occurs early in the incident, so that support is in place and available when it's needed. After a period of time, the intensity of being in command can cause even experienced, tough ICs to become somewhat catatonic and lose their focus. The folks who come in to relieve and support command become (in effect) the IC's strategic-level RIC team.

Safety Effect:

They (the relief command team) also arrive with "new eyes." Sometimes, everyone who has been on the scene for a long time has just gotten used to looking at dangerous conditions. When the new arrivers look at such conditions, they do not have the same perspective, so they can use their new evaluation to adjust the attack plan to match current conditions. The haz-zone workers richly deserve to be protected by an adequately staffed command team that is continuously awake, lucid, and responsive.

Command Safety

Continue/Transfer/Terminate Command

IC Checklist:

- ☐ Assume, maintain, and upgrade effective command positioning.

Safety Effect:

In most cases, local incident operations will start off with a fast-attacking IC (the company officer of the first unit on the scene). This is a good thing. A fast-attacking, initial-arriving company officer IC can effectively assign their company and the next several units, and personally direct putting a hit on the incident problem (while it is the smallest). This provides enough fast, front-end organization to get the initial attack wave up and going. If the initial attack efforts do not stabilize the problem, command should be transferred to an IC who will be operating in the command mode (remote, stationary, and supportable position), as quickly as possible. The new IC is in a better physical position to expand the operation and keep track of where everyone is, what they are doing, and if they are okay, which is a critical command safety function. A major command and safety system capability involves a tiered level of regular response, where a fast-attacking team of initial arrivers can go to work quickly and be confident that responder's bosses, who automatically come in behind them, will use SOPs to transfer command, and will assume a more "sterile" command position. A major objective of this integrated system is that both command and control and incident safety is continually strengthened. This reinforced support occurs when the IC uses the CP advantages to

Safety Effect:

create a command team (sitting bosses) and makes sector assignments that put walking bosses close to working hazard-zone companies, who are commanded by working bosses. This regular, decentralized, integrated (sitting/walking/working bosses) approach automatically reinforces and supports the IC's capability to stay inside the CP and act as the strategic manager of the overall incident (i.e., sitting boss). It's hard to do this when the IC is standing outside in the dark, getting rained on, and the wind just blew away the TWS with all the incident assignments on it, and six very excited responders are screaming directly in their ear.

Command Safety

Continue/Transfer/Terminate Command

IC Checklist:

☐ Develop and maintain effective fireground communications.

Safety Effect:

The fast-attacking IC is generally in a horrible interior communications environment. Generally, that IC is bundled up in a PBI flashover monkey suit, looking through a fogged-up SCBA face piece, and crawling around in the products of combustion (very exciting!). When command is transferred (upgraded) to an IC in the command mode (i.e., positive), the communications capability of command skyrockets. To understand the effect of this upgrade, just listen (as you monitor the radio) to how different IC #2 sounds in the CP over the tactical channel than IC #1 sounded from an inside attack position. The new IC is in a better position to assign tactical-level supervision sector responsibilities (if they haven't been assigned already). Establishing sectors creates decentralized geographic and functional command partners, which will greatly increase the IC's capability to exchange information from all over the incident site.

Command Safety

Continue/Transfer/Terminate Command

Safety Effect:

Another major safety factor involving locking the IC up in the CP creates the capability for the IC to be continuously available to effectively receive, process, and respond to (quickly answer) hazard-zone communication. This involves maintaining, managing a current inventory, and tracking of who is working in the hot zone, so the IC knows who they are commoing with, critically listening to any commo coming into the CP, and being able to answer that commo on the first transmission into the CP from a hazard zoner. The physical position of those operating on the three levels (task--companies/tactical--sectors/strategic--IC) creates a commo capability that is a direct reflection of the typical conditions that go with that position. Task-level operating companies are mostly in tough spots/ sector officers are a little bit better because they are not distracted by the competition of doing physical work along with commo/the IC in the CP can use the advantage of that CP to "make up" for the difficult positions of the other two levels. Effective safety and effective communications are inseparable... simply, when commo turns to "doo-doo," so does command and control, along with the welfare of the troops.

Note:

We have exhausted the reader with the recurring importance of the IC continually

Command Safety

Continue/Transfer/Terminate Command

Safety Effect:

being able to control the position and function of those working in the hazard zone. This critical capability depends directly on the actual commo connection between the IC and the responders who are inside the hot zone. The IC uses the standard IC commo system (SOPs, techniques, application) to get the workers inside to maintain an ongoing awareness of their welfare and to quickly get them out if necessary. We cannot over emphasize how important commo is to safety. We must think of the IC as the main commo player, and the commo system as the basic inside/outside safety connection ("umbilical cord").

Command Safety

Continue/Transfer/Terminate Command

IC Checklist:

☐ Keep the attack plan going.

Safety Effect:

The IC must realize two basic truths:

1. When the fire goes out, everything gets better (and safer).

2. The fire always goes out (eventually).

The major survival factor contained in these realities requires the IC to make a strategic (offensive/defensive) decision about where the attack will occur (operationally) based on an evaluation of the critical factors on current and forecasted conditions. While this is very simple (and very challenging), it is critical that the IC continually press the attack to solve the basic incident problem. How (and where) this "press the attack" will occur must be packaged up in the overall strategy and the related IAP. "Press the attack" means the IC must (in terms of operational "force") create enough of an operational response (big enough, fast

Safety Effect:

enough, in the right place) to pull off the objective of the selected strategy:

Offensive: Go inside, search and rescue, protect and maintain in/out path, directly attack fire with adequate water, provide support (forcible entry/ventilation/access provision), continue evaluation, stay inside as long as standard safety system will protect the troops--if safety score goes negative... get out, go defensive. IC must always consider and react to the ongoing capability for inside offensive attackers to maintain inside tenability and to always be able to exit the hazard zone.

Offensive/Marginal: Go inside, protect and maintain in/out path, provide whatever attack is required to protect searchers, get an "all clear"--evaluate conditions: if they are getting better, continue attack (see offensive routine); if they are getting worse... drag ass--go defensive (same routine about ongoing exit capability: NO EXIT = defensive).

Command Safety

Continue/Transfer/Terminate Command

Safety Effect:

Defensive: When the combination of fire extent and severity, collapse potential, and arrangement exceeds standard safety system capability = defensive. Get out, stay out, stay away from collapse/hazard zone.
Defensive conditions require us to stay out of collapse/hazard zone, control positions, surround and drown, protect exposures, eat Snickers bars, go home okay... tell owner to sprinkler next time.

Once the IC gets the attack going, those efforts must be continued (and escalated, if necessary) until a standard outcome is achieved. The IC must deploy around the basic (size/time) event profile. This requires continuing to escalate (call for/assign/manage) additional resources to match that profile. Be very careful of having the fire exceed the attack/operational capability, simply because it burns past the level of response that generally controls fires in that local area. Most of our response systems are excellent two-attack line departments that fight mostly house fires. If those two lines don't quickly control the problem, we are many times almost disoriented and unable to escalate beyond those two lines. This creates a "one-trick pony" effect that can seriously expose the hazard-zone troops who are now

Command Safety

Continue/Transfer/Terminate Command

Safety Effect:

physically overmatched on the two fast-attack lines (that generally are effective locally) when the fire keeps burning (i.e., that "trick" doesn't work). An example of this process occurs when we take single-family attack routines into commercial fires. If, for whatever reason, the attack effort runs out of gas before the fire, the fire wins. Fires are bad-ass events that injure, kill/damage, and destroy everybody and everything that they get on. The horrible effect of being up close (and personal) to those fire conditions occurs more often to our troops than any other single group of people. Regardless of how many hose lines the homeys are good at using, they will eventually meet up with a fire situation that exceeds that familiar and comfortable capability. When this happens, the IC (and everyone else) must recognize they are essentially overmatched and then the only strategy (and tactical IAP) is to create operating positions that do not get those defensive conditions on the hazard zoners. This is not a failure, it is just life in the street. Most of our local incidents are offensive, but sometimes defensive occurs in the middle of changing offensive times--other times, defensive is waiting for us... the guy who said something about "live another day" was a smart (old) dude.

Command Safety

CONTINUE/TRANSFER/TERMINATE COMMAND

IC Checklist:

☐ Use standard command transfer (both ways).

Safety Effect:

Hazard-zone workers must be continually protected by an in-place IC (and an adequate command team/staff). Big-time safety problems occur when there is no IC, or when there are multiple competing people on the scene who each think (and act) like they are the IC. Standard command transfer is one of the ways the command system insures there is just one IC operating at a time. This procedure is critical because it describes the local details of how command will actually be passed from the donor to the recipient, as ranking officers arrive on the scene. The reason we transfer command is to upgrade and strengthen the IC function, as higher-ranking, better-qualified, older, more experienced, better-positioned and equipped officers arrive on the scene. It should occur for functional improvement, not for ceremonial (ego-based) reasons. Having multiple, ranking officers present on the scene can be very confusing (and unsafe), if such a standard command transfer process is not agreed upon and used. The ranking officer who is going to take command should quietly monitor the tactical channel while responding, and note the assignments and action that have taken place before their arrival. When they get on the scene, they should go to the current IC (if possible) and do a face-to-face transfer. The standard command transfer questions

Command Safety

Continue/Transfer/Terminate Command

Safety Effect:

are: "What do you have?"; "What are you doing?"; "What's your plan?"; "What do you need?"--followed by, "I've got it." This means the new IC now is basically oriented to the situation status and has taken command. If the current IC is in an operational position (particularly inside the hazard zone as a fast-attacking initial company officer IC), the transferring officer can contact the IC and transfer command over the radio. That sounds something like, "I have copied what has occurred on the radio--I'll take it out here." In the old days, we would transfer command for the first forty-five minutes of the incident to every subsequent arriving officer who outranked the current IC. In those days, every boss would retain their "car number," so everyone had to struggle to maintain a moment-to-moment awareness of who was the current incident boss. This created a chain-of-command form of "musical-chairs" (i.e., chiefs) response and made it difficult to figure out who was actually the IC. A lot of times, not only the IC changed, but so did the general plan (IAP), and the overall approach with each newly-arriving IC, based on the tactical "personality" that every individual brought to the operational period when they were in command. Command transfer should not be an endless process and there should be a limit to how many times it should occur. Generally, command will be passed from the initial-arriving company officer IC to the first-arriving response chief. Once a

Command Safety

Continue/Transfer/Terminate Command

Safety Effect:

command officer is in place as the IC, then more arriving chiefs should be assigned to building the standard command team, with the subsequent arrivers serving as the SO and SA. Building the command team is the major (and most effective) way we strengthen command and take advantage of more command "horsepower" being present. The command team creates the structure where such capability can be actually mobilized, integrated, and applied to providing incident command and support. Being able to quickly escalate our continuing command capability to be able to match an expanding incident is a major way we continuously protect the troops.

Another major capability of the command team is that they are able to reinforce, expand, and provide support to the IAP. This creates the very practical ability to be able to quickly achieve the tactical priorities in offensive places, and within offensive periods. A lot of command-team capability is simply a function of collective human dynamics (physics?). Three people who are each performing as a command-team member, where every person plays a complimentary and integrated role, can outperform a single person who is operating by themselves--while it's not complicated, the synergistic effect is very powerful. While the command team (itself) does not create any physical effort or outcome, it can call for, order into place, and effectively manage more resource than a single person. The team can also maintain a

Safety Effect:

level of ongoing control over the position and function of hazard-zone workers more effectively than a single IC, and such control always increases the safety of those workers.

We also create another major safety capability, as we standardize our operational procedures, and the effect these procedures create. As we refine SOPs, tactical guidelines, and standardize IMS, we begin to develop a regular way of doing business--this causes the players to become highly interchangeable. While older and more experienced members generally bring more capability (more "slides" in their tray) to the command process, it's not because they implement secret procedures that only they know, it's simply because they have seen more actual tactical situations, where the widely understood SOPs were applied. They also have more experience in how to handle themselves and have had more time to figure out how to personally act effectively. There is no substitute for operational experience, and we should take advantage of it by using it when we have it, and then teaching the lessons the old probably learned the hard way to the young, so it's not quite so painful to get old.

Command Safety

Continue/Transfer/Terminate Command

IC Checklist:

- ☐ All pertinent information must be exchanged between IC and company/sector officers.

Safety Effect:

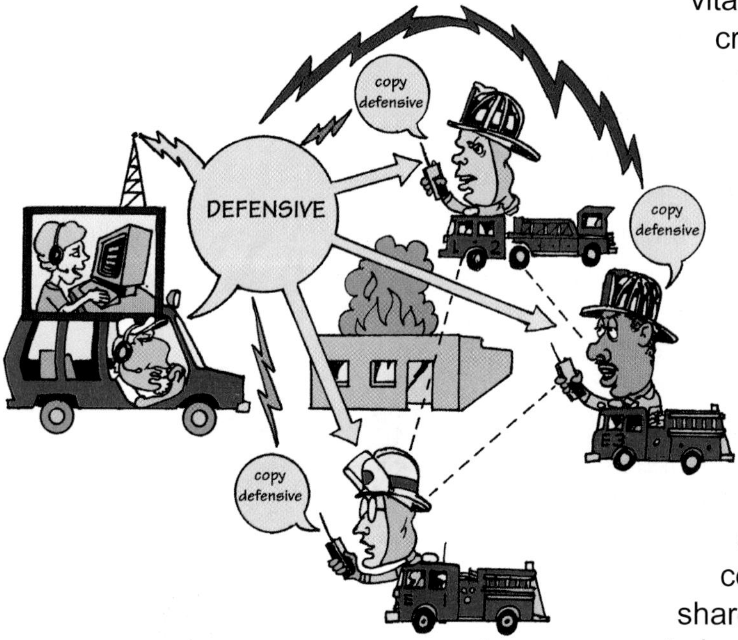

A major way the IC creates and maintains continuous, safe operations is by setting up and using an effective incident organization. Every standard organizational piece and part manages operations in their area/ function and exchanges information about what is going on in their location. This is vital for overall scene safety. Any critical factor that is going to affect the IAP, completion of the tactical priorities, or most importantly firefighter safety, must be shared with the effected parties (sounds like lawyer talk). Some standard safety benchmarks should be built into the system. Things like PARs, RIC team readiness, and regular safety updates from safety sector officers or companies need to be transmitted to the IC. When the IC gets a report that could impact firefighter safety in certain (or all) sectors, this must be shared with the appropriate people, and get plugged into the IAP. Those safety changes must be quickly communicated to (and reverified by) the haz zoners. The launching pad for this information exchange is a common understanding throughout the team of both the capability of our basic safety system and an awareness of how the regular and special incident hazards (remember the gauges) evolve, behave, and

Safety Effect:

affect our humans, in relation to how well the safety system will protect the troops in that particular situation. Effectively applying this standard safety approach is a big deal and requires that everyone know the details of how the basic "safety score" (relationship of incident hazard to safety system capability) will be kept, and how our SOPs are used as an organizational and safety game plan. That SOP-based plan creates the regular organizational formations and the operational moves we use to get ahead of (and in some cases we use to stay out of the way of) the incident problem(s). Effectively exchanging this information is the only way the IC can make the continuous changes required to match evolving conditions. This organizational agility involves making such transitions like reinforcing (making bigger) attack positions, coordinating and integrating multiple functions performed by different teams, shifting tactical positions (i.e., moving the troops) to cut off and control shifting/expanding fire conditions. Being "under control" doesn't mean that we are always able to control incident conditions. What it does mean is that we are in control of the position and function of our responders and resources in relation to those conditions. During defensive operations, the fire is simply too big, spread out, or complicated for our resources to overpower. When this happens, we set up outside and away from the hazard zone, protect our troops, and let the fire burn down to the size of our control capability. Then we put out what's left--what defensive means is that the fire is out of

Safety Effect:

control in relation to the size of our operational capability. A huge safety problem always exists when we are out of control, regardless of the size of the incident problem. Lots of firefighters are injured and killed at nickel-and-dime fires in small fire areas, simply because they (the firefighters) are out of control. The regular incident hazards are always present in these situations and those hazards hurt/kill them. What being out of control from a safety standpoint means is that some part of the standard safety system is not effectively in place, and this system deficiency exposes a firefighter to the incident hazards. We are in control of the pieces and parts of our standard safety system when based on that organizational behavior, the safety system components are either in place or not in place. This is the part of the safety equation we control. Everyone must be able to operate with the confidence that every team member, on every level, is completely performing their safety role. Once this level of understanding and actual compliance is in place, then we can go out and play in the exciting world of strategy and IA planning because our safety system is effectively under control. A huge part of the IC's safety routine must involve quick, short, and current information exchange about incident conditions. If the IC cannot make an ongoing series of tactical shifts to match how the event is changing

Safety Effect:

(hazards vs. safety system) based on accurate information, the troops can become quickly threatened by those conditions. This is what being out of control from an operational standpoint really means. Out of control and unsafe pretty much always go together.

Command Safety

Continue/Transfer/Terminate Command

IC Checklist:

☐ Place resources back in service with a demobilization plan.

Safety Effect:

At the end of incident operations, the command system must develop a demobilization plan designed to send everyone who worked at the event home in an organized way. The IC must always keep enough command (people and organization) in place to manage the level of operations that are present. Reversing the same TWS resource inventory tracking system the IC used to put resources to work is an effective way to focus on who is on the scene, where they are, and how "releasable" they are. It makes sense (as much as possible) to release resources based on their order of arrival. The responders who got there first are the most fatigued and should be accounted for, rehabbed, checked by medics, thanked, and sent home first. Subsequent arrivers should "stay in line" and go through the same standard work cycle. Bringing in fresh troops and rotating companies to avoid excessively beating up our humans makes sense. It's lots smarter to send the troops home while they are still ambulatory, basically sane ("B" shift?), with all their body parts intact, than to wear them down to a point where they must go from the incident to intensive care (duh!). In the old days, the dinosaur bosses used to say about the first arrivers (who got there

Safety Effect:

at midnight), "It's your fire," and they would get to see the sun come up with most of the troops laying in the street with their heads resting on the curb... very stupid way to manage what we say is our most important asset. The IC should give the same attention to winding down that was given to winding up.

Command Safety

Continue/Transfer/Terminate Command

IC Checklist:

☐ Insure adequate critique process is underway before decommitment.

Safety Effect:

Our local experience is the most important test of our command and operational system. Where and when River City resources (humans and stuff) and River City systems (SOPs) must go to work on an actual River City situation (the Smith family fire) becomes show time for all our procedures, training practices, and refinement. After an incident, we should carefully examine how our people and procedures performed by conducting a standard, post-incident critique. The standard critique format includes:

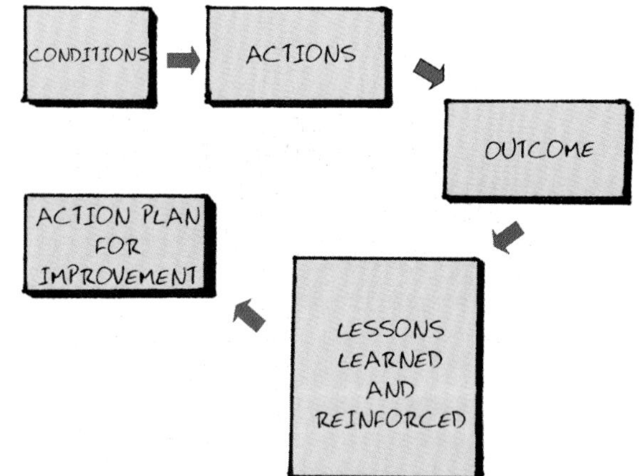

- conditions
- action
- outcome
- lessons learned and reinforced
- action plan for improvement.

How the safety of responders was managed should be a major critique component. Actual injuries (or worse) should be thoroughly investigated and discussed. Near misses should be openly discussed and treated as if they were hits. How the safety "score" occurred and was managed throughout the incident should be outlined along with any lessons that were learned. Using a series of "snapshots" of the stages of how the incident evolved, and then describing the command operational and safety response in

Command Safety

Continue/Transfer/Terminate Command

Safety Effect:

each stage, is a useful way for everyone to understand how the beginning, middle, and end actually occurred. This approach can also describe how those stages connected (or disconnected) to/with each other, and how the team made those transitions from one stage to the next. Lots of times, those transitional times can be very dangerous, so it becomes critical that we develop the ability to create seamless, streamlined, safe changes to match shifting conditions, and to protect the troops during those periods. Critiques should be a regular part of every incident, and based on the size and complexity of the event, the critique should be small--"tailboard"--station/medium--battalion/large--department wide. The critique should be done as soon after the event as possible, and should follow a standard format that is practiced, so everyone can expect (and depend on) how the process will occur. The critique should be straight forward and honest in dealing with what actually happened. If we did well--we should say so, and we should support such performance by listing those positive experiences as "lessons reinforced." If we screwed up, we should outline what happened, what we would do differently, and should list that experience as a "lesson learned." The point of the critique is to use our actual experience to improve future performance. The action plan for improvement becomes the way that we package up that process. The improvement plan should describe in detail who is going to do what,

Command Safety

Continue/Transfer/Terminate Command

Safety Effect:

by when, to implement the lessons learned. The whole process should be humane and survivable, with no heads left on sticks. We will never get better (or safer) if we don't honestly and critically examine our command/operational/safety performance--this involves commending and reinforcing standard performance and coaching through problems. This is a big-deal boss responsibility.

Command Safety

Continue/Transfer/Terminate Command

IC Checklist:

☐ Provide required critical incident support.

Safety Effect:

Firefighters are called upon (very typically) to do some customer service stuff under really difficult conditions. They are the local responders of first and last choice. If the customer's kitchen is on fire, we are called first--if there is loneliness, fear, despair, and no one else will come, we are called last (because we always respond). Our humans are present at birth, death, injury, insult, accident, and hallucination. Many times, they can prevent harm, but sometimes that harm has already occurred, and they must deal with what is left. The IC must always attend to the physical needs of our troops... provide rehab stuff like food, fluid, rest, cooling/heating, and Snicker bars. Other times, the IC must also provide support for the feelings (and the effect of those feelings) that emerge from the sad part of our business that is heart breaking. Such emotional support can range from a simple debriefing to a full critical incident intervention. Lots of times, just having a BC stop by the station with a dozen donuts and personally and informally discuss the incident fits the typical non-ceremonial firefighter personality. The power of this approach is not so much that the boss says

Safety Effect:

something that is magical and then everything is okay. It works because the boss is present with the troops when they (including the boss) are feeling really badly about something in which they were all involved. Simply, most firefighters can relate a lot better to their regular boss than a team of psychotherapists. Sometimes, after the boss deals with E-1, they decide the problem exceeds donut therapy and they send some of them, or all of them (including the chief), for a clinical experience. The most important, critical support thing in a fire department is how we treat each other every day--this sets the stage for what happens after a difficult experience... what happens to us two weeks after a difficult event is a direct reflection of what was going on inside our organization two weeks before that event. A huge element in the troops recovering from difficult incidents is the feeling that the event was well managed and that everyone on the team did everything humanly possible to resolve the problem. No matter how much they want to, or how hard they try, firefighters can only save the savable. Based on these feelings, IMS should always create an experience where being sad with the outcome (mostly to humans) is not because the system was not in place, or that it didn't work as it was designed... simply, well-managed incidents conducted inside a well-managed organization become highly therapeutic to the players.

Command Safety

Continue/Transfer/Terminate Command

This happens when Continue/Transfer/Terminate Command is done:

- ☐ Initial command is assumed (so there is something to transfer).
- ☐ Command transfer is used to improve, upgrade, and strengthen command (functions).
- ☐ We always have a single IC (one at a time).
- ☐ We effectively use our individual and collective command capability (experience, rank, speciality, etc.).
- ☐ The initial size up is reviewed and adjustments can be made if necessary.
- ☐ Resources are ordered and deployed to match the incident forecast.
- ☐ We assemble and maintain the resources to outlast the incident problem (and kick its ass).
- ☐ Our safety plan matches our forecast.
- ☐ We learn from the seasoned, and appreciate and teach the young.
- ☐ We support/assist/encourage/enjoy each other every day.
- ☐ We help each other recover from difficult experiences.
- ☐ We/customer win.

Command Safety

Continue/Transfer/Terminate Command

This happens when Continue/Transfer/Terminate Command is not done:

- ☐ Big mess occurs when IC #1 is in place, multiple bosses arrive, they all start to yell/shout--none of them formally transfers command--or some hyperventilating chief takes command two miles away.
- ☐ No one knows who/where IC is--troops hide out.
- ☐ Non-unity of command radio commo gets jammed up with multiple out-of-control bosses (all blabbing).
- ☐ No one does an effective event profile (particularly a forecast of duration).
- ☐ We fail to expand and strengthen command.
- ☐ We get behind, play catch up--never catch and overtake the power curve.
- ☐ We run out of command while the incident problem(s) go on.
- ☐ We get angry (and dumb) with each other.
- ☐ We/customer lose--problem wins.

Command Safety

Continue/Transfer/Terminate Command

COACHING VERSION

"Set up command to outlast the fire. Use the CP advantage, establish sectors, invest in sound, current evaluation, connect everyone with effective communications. Use the command staff to support long-term IC impact and durability. Don't beat up the entire command staff during the front end of long campaign events. Forecast incident time and assign rotating teams to provide ongoing command. For extended operations, develop a series of ongoing IAPs that provide the planning and timing to get through the event. Always set up and operate in a way that is "transferrable" to your boss. Effective command transfer and escalation uses organizational management capability to strengthen, expand, and improve command. All the fire understands is water and could care less if a captain or deputy chief is the IC. If you can't improve the quality of command, simply don't transfer it. Check in with the troops as operations wind down. Be certain they are all in one piece mentally and physically. Provide whatever resources and attention that are required to help them. Start the critique process before everyone goes home. Let the troops describe the action they engaged in and make any suggestions they have to improve performance. Use the experience of this incident as the basis for future training, practice, and procedure refinement. The IC consciously "loads" the lessons of this event into your "mental files" for future reference. This approach creates the very practical capability to use experience as the foundation for growing old gracefully (i.e., smarter, nicer)."

Command Safety

Continue/Transfer/Terminate Command

Command Safety

Glossary of Acronyms

AC air conditioner
AED automatic external defibrillator
AVB Alan V. Brunacini
AVL automatic vehicle locator
BC battalion chief
BRT big red truck
BTU British thermal unit
CB citizen's band
CF cluster_ _ _ _
CP command post
DRT dead right there
EMS emergency medical services
EOC emergency operations center
ET elapsed time
FAA Federal Aviation Administration
GPM gallons per minute
GPS global positioning system
IAP incident action plan
ICMA International City Managers Association
IC incident commander
ICS incident command system
IFSTA International Fire Service Training Association
IMS incident management system
IRIC initial rapid intervention crew
IRR initial radio report
MBA Master of Business Administration
MDT mobile digital terminal
NFPA National Fire Protection Association
NTSB National Transportation Safety Board
OD on deck
OPS operations
PAR personnel accountability report
PBI polybenzimidazole (protective clothing material)
PC personal computer
PFD Phoenix Fire Department
PIO public information officer
POC products of combustion
PPE personal protective equipment
RIC rapid intervention crew (aka RAT, RIT, etc.)
RNB Robert Nick Brunacini
RTO recruit training officer
SA senior advisor
SCBA self-contained breathing apparatus
SO support officer
SOP standard operating procedure
SU set up
SUV sport utility vehicle
TIC thermal imaging camera
TWS tactical work sheet
ZIP zero impact period

Command Safety

Glossary of Terms/Abbreviations

Ack	acknowledge
Blue Card	IMS designed to manage active, fast, local-level incidents which require an immediate incident commander at the scene in control during incident escalation.
Commo	communications
Emergency Traffic	clears the tactical frequency of routine radio traffic; gets a special tone; reports critical safety conditions (i.e., Maydays)
Harden	protect, reinforce, add resources to increase safety
Mayday	signals that a firefighter is in trouble and requires rescue
Ops	operations
Recon	reconnaissance
Red Card	IMS designed to manage large-scale, long-term (mostly wildland) incidents
Tactical Priorities	Safety, rescue, fire control, property conservation, customer stabilization which are basic operational priorities

Other Books by Author

Fire Command
Copyright 1985
available through NFPA 1-800-344-3555

In *Fire Command*, Fire Chief Alan Brunacini shares his expertise on everything from basic fire fighting techniques to tactics of command. You'll sharpen your fireground skills by learning proven techniques for developing attack plan, placing apparatus, conducting rescue operations, managing fire streams, and other critical functions. Each chapter provides clear objectives--plus illustrations, examples and a report card for self evaluation.

Essentials of Fire Department Customer Service
Copyright 1996
available through IFSTA 1-800-654-4055

Essentials of Fire Department Customer Service is a common-sense manual on putting the service back in the fire service and written in a humorous conversational style for those who work with the public.

***Fire Command,* 2nd Edition**
Copyright 2002
available through IFSTA 1-800-654-4055

Fire Command, 2nd Edition, is based on Fire Command original content but includes fifteen years of additional fireground experience which makes this 2nd edition a more powerful and relevant book for today's firefighters. Using his characteristically engaging and street-smart style, the Chief mixes hard-hitting practical information and strategies with humorous anecdotes to present a structured approach to incident management that is easy to relate to and understand. Develop know-how, confidence, and the ability to adjust to changing situational demands on the fireground. This book devotes a chapter to each of the eight command functions, with a description of how the incident commander performs that function and the part it plays in the total incident management system.

This book has a companion workbook, study guide, and a nine-set fire command video series which add an additional dimension to fire command instruction.

Timeless Tactical Truths
Copyright 2003
available through IFSTA 1-800-654-4055

Timeless Tactical Truths is a small book with cartoons which offers readers a collection Bruno's most insightful one liners--observations he has made during his 45 years in the fire service. Many of these one liners have become "Brunoisms" out in the world.

Soon to be Published Books by Author

Anatomy and Physiology of Leadership
Tactics and Strategy

Command Safety